NEUTRON SCATTERING IN BIOLOGY, CHEMISTRY AND PHYSICS

NEUTRON SCATTERING IN BIOLOGY, CHEMISTRY AND PHYSICS

A ROYAL SOCIETY DISCUSSION

ORGANIZED BY
SIR RONALD MASON, F.R.S., E. W. J. MITCHELL
AND J. W. WHITE

HELD ON 26 AND 27 SEPTEMBER 1979

QC793.5
N4628
N5
1980

LONDON
THE ROYAL SOCIETY
1980

Printed in Great Britain for the Royal Society
at the
University Press, Cambridge

ISBN 0 85403 151 0

First published in *Philosophical Transactions of the Royal Society of London*,
series B, volume 290 (no. 1043), pages 479–681

Published by the Royal Society
6 Carlton House Terrace, London SW1Y 5AG

PREFACE

Neutron diffraction experiments had their origin just over 30 years ago; wider aspects of neutron scattering and neutron spectroscopy are more recent innovations, at least at the present level of resolution and sensitivity. High flux sources are having a fundamental impact on the scope and value of experiments and it is expected that further significant advances will be made with pulse sources and new instrumentation, including novel signal and data processing methods. In this Discussion Meeting, therefore, we wanted to take stock of the present position but in the spirit of providing a preview of likely developments for the future, rather than a retrospective view of the major accomplishments of the past.

In structural crystallography we have tended to highlight the applications of neutron diffraction to the investigations of large molecular structures and to spin density distributions. Diffuse scattering is related to defects and magnetic properties, while views on low-order systems are reflected in papers on adlayer structures on surfaces, on aqueous solutions and on liquid crystals. Some of the advances in solid state physics are covered by discussions of antiferromagnetic insulators and incommensurate structures, while the scope of neutron scattering in fundamental particle physics is shown by contributions to our understanding of elementary excitations in liquid helium-3 and the recent high accuracy determination of the electric and magnetic moments of the neutron.

It was essential to have this record of the Discussion Meeting available as early as possible. We are grateful to all the contributors for the high standard and originality of the papers and the way that they have helped to make publication a relatively rapid one.

July 1980

RONALD MASON
E. W. J. MITCHELL
J. W. WHITE

CONTENTS

P. Jane Brown, J. B. Forsyth and R. Mason, F.R.S.
   Magnetization densities and electronic states in crystals — [1]

M. Leslie, G. T. Jenkin, J. B. Hayter, J. W. White, S. Cox and G. Warner
   Precise location of hydrogen atoms in complicated structures by diffraction of polarized neutrons from dynamically polarized nuclei — [17]

G. A. Bentley and S. A. Mason
   Neutron diffraction studies of proteins — [25]

E. W. J. Mitchell and R. J. Stewart
   Diffuse neutron scattering from crystal imperfections — [31]
     Discussion: R. J. R. Miller, D. J. Cebula — [46]

W. Schmatz
   Magnetic diffuse and small-angle scattering — [47]

G. Bomchil, A. Hüller, T. Rayment, S. J. Roser, M. V. Smalley, R. K. Thomas and J. W. White
   The structure and dynamics of methane adsorbed on graphite — [57]
     Discussion: A. D. Buckingham, F.R.S. — [72]

J. E. Enderby
   Neutron diffraction, isotopic substitution and the structure of aqueous solutions — [73]
     Discussion: E. W. J. Mitchell, J. G. Powles — [85]

J. C. Frost, A. J. Leadbetter and R. M. Richardson
   Molecular crystals and liquid crystals: new results for t-butyl chloride — [87]

R. A. Cowley, F.R.S.
   Percolation in antiferromagnetic insulators — [103]

J. D. Axe
   Incommensurate structures — [113]

K. Sköld and C. A. Pelizzari
   Elementary excitations in liquid ^{3}He — [125]

J. M. Pendlebury and K. Smith
   The electric and magnetic moments of the neutron — [137]

B. Jacrot
   The use of neutrons to study protein–RNA interactions — [147]

J. T. Finch, A. Lewit-Bentley, G. A. Bentley, M. Roth and
P. A. Timmins
Neutron diffraction from crystals of nucleosome core particles [155]

H. D. Middendorf and Sir John Randall, F.R.S.
Molecular dynamics of hydrated proteins [159]
Discussion: A. J. Leadbetter [174]

B. E. F. Fender, L. C. W. Hobbis and G. Manning
The U.K. Spallation Neutron Source [177]

T. Springer
Developments in experimental neutron physics at the Institut Laue–Langevin [193]

Magnetization densities and electronic states in crystals

By P. Jane Brown†, J. B. Forsyth‡ and R. Mason§, F.R.S.

† *Institut Max von Laue – Paul Langevin, Avenue des Martyrs, 38042 Grenoble cedex, France.*
‡ *Science Research Council, Rutherford Laboratory, Chilton, Didcot, Oxon. OX11 0QX, U.K.*
§ *Now Sir Ronald Mason, School of Molecular Sciences, Sussex University, Falmer, Brighton, Sussex, U.K.*

The classical polarized neutron technique provides an extremely sensitive method for studying magnetization distributions in crystalline materials. In the transition metals and their compounds it is recognized that the d electrons act both as valence electrons and as carriers of the magnetism. This dual role implies that the magnetization distribution can give information about the behaviour of valence electrons. The pioneering work in this field yielded new insight into the behaviour of the magnetic elements themselves.

The paper begins with an introduction to the elastic magnetic scattering of neutrons, the electronic origin of magnetization density and the polarized neutron technique itself. A brief survey of earlier work in the important areas of application is followed by more detailed discussion of three recent experiments: the determination of the paramagnetic form factor of technetium, a study of orbital effects in a ferrimagnetic vanadium salt ($K_5V_3F_{14}$) and the spin density and bonding in the $[CoCl_4]^{2-}$ ion.

1. Introduction

At the start of this meeting it is perhaps not out of place to recall some of the properties of the neutron. It is a neutral particle with mass not very different from that of the hydrogen atom; it has a magnetic dipole moment of 1.9 nuclear magnetons but to a close approximation no electric dipole moment. The neutron interacts with atomic nuclei through the strong nuclear interaction, and with electrons by virtue of its magnetic moment. It is this second interaction that is of prime interest for this paper. Since the neutron is scattered by its magnetic interaction with electrons, the measurement of magnetic scattering enables information about the magnetic properties of solids and in particular the distribution of electronic magnetization to be determined.

Within the first Born approximation, elastic scattering cross sections may be written in terms of the Fourier transform of the interaction potential. Thus

$$d\sigma/d\omega \propto \int \psi^* V(r) e^{ik \cdot r} \psi \, dr^3. \tag{1.1}$$

For neutron magnetic scattering the interaction potential is related directly to the magnetic flux density, B, in the scatterer and through that by Gauss's theorem to the magnetization density $M(r)$ defined as the magnetic moment per unit volume. A magnetic structure factor $M(K)$ can be defined such that

$$M(K) = \int M(r) e^{ik \cdot r} dr^3. \tag{1.2}$$

Then the magnetic elastic scattering cross section for unpolarized neutrons is given by

$$\partial\sigma/\partial\omega \propto |\hat{k} \times M(k) \times \hat{k}|^2 = |M_\perp(k)|^2. \tag{1.3}$$

$M_\perp(k)$ is the generalized magnetic interaction vector and it is this quantity, its square or its components that can be measured in a magnetic scattering experiment.

2. The electronic origin of the magnetization density

The magnetization density in a crystal is due both to the intrinsic magnetic moment of electrons (the electron spin) and to the magnetic moment generated by moving electrons (orbital magnetization). In most materials, paired electrons generate equal and opposite magnetic moments so that the net magnetization density is everywhere zero, but in magnetic or magnetized matter some electrons are unpaired. In such cases the magnetization density due to spin reflects the spatial distribution of the unpaired electrons, but that due to their orbital motion is less simply interpreted.

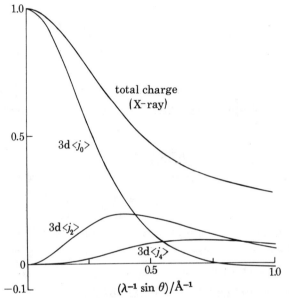

Figure 1. The total charge and 3d form factors for the Mn^{2+} ion.

In some cases it may be appropriate to consider the total magnetization as the sum of distributions from individual atoms, and it is then useful to define magnetic form factors for each atom. A magnetic form factor is defined as

$$f(k) = (\int m(r) \, e^{ik \cdot r} \, dr^3) / (\int m(r) \, dr^3), \qquad (2.1)$$

where $m(r)$ is the magnetization density associated with the atom. The magnetic structure factor can readily be expressed in terms of the form factors as

$$M(k) = \sum_n f_n(k) \, M_n \, e^{ik \cdot r_n}. \qquad (2.2)$$

Here $f_n(k)$ is the form factor for the nth atom at distance r_n from the origin, and M_n is its magnetic moment; the sum is over all atoms in the crystal. When the magnetization is due to electrons in a single unfilled shell, the form factor may be written as

$$f(\boldsymbol{k}) = \sum_l A_l(\hat{\boldsymbol{k}}) \langle j_l(|\boldsymbol{k}|)\rangle, \qquad (2.3)$$

where

$$\langle j_l(\boldsymbol{k})\rangle = \int_0^\infty U^2(r) j_l(kr) \, dr. \qquad (2.4)$$

Here $j_l(kr)$ is the spherical Bessel function of order l and $U^2(r)$ the radial distribution function of electrons in the open shell. The $A_l(\hat{\boldsymbol{k}})$ are coefficients that depend both on the direction of the scattering vector, $\boldsymbol{k}$, and on the magnetic configuration. For a spherically symmetric spin-only ion such as the Mn^{2+} ^{5}d configuration, only A_0 is finite. For the general spin-only case with d electrons, A_0, A_2 and A_4 are non-zero. An approximation to the spherical form factor valid for small $\boldsymbol{k}$ when orbital moment is present is given by the dipole approximation (Marshall & Lovesey 1971)

$$f(\boldsymbol{k}) = 2S\langle j_0(k)\rangle + L(\langle j_0(k)\rangle + \langle j_2(k)\rangle). \qquad (2.5)$$

Figure 1 shows the k dependence of the three integrals $\langle j_0\rangle$ $\langle j_2\rangle$ $\langle j_4\rangle$ for 3d electrons in the Mn^{2+} ion. Only $\langle j_0\rangle$ is non-zero at $k = 0$, so the coefficient A_0 measures the net magnetic moment of the ion. Figure 1 also shows the X-ray form factor for Mn^{2+} to which all the electrons contribute; the contribution from the core electrons makes this form factor larger at high k than the magnetic form factor. One may compare the absolute magnitudes of the scattering factors at $\lambda^{-1}\sin\theta = \sin\theta/\lambda = 0.25$ Å^{-1}[†] for X-rays and for magnetic and nuclear scattering of neutrons; these are 5.5, 0.74 and -0.37×10^{-12} cm respectively.

3. Scattering of polarized neutrons

The scattering cross section of (1.3) is the sum of two partial cross sections that correspond to scattering with and without a change of neutron spin direction. There can be interference between nuclear and magnetic scattering if there is no change of spin direction and for an incident beam polarized parallel to $\hat{\boldsymbol{\lambda}}$, the spin-non-flip cross section is,

$$\partial\sigma/\partial\omega(\uparrow\uparrow) = |N(\boldsymbol{k}) + \hat{\boldsymbol{\lambda}}\cdot\boldsymbol{M}_\perp(\boldsymbol{k})|^2. \qquad (3.1)$$

Only magnetic scattering can reverse the neutron spin direction in Bragg scattering if there is no nuclear polarization and the spin flip cross section is simply

$$\partial\sigma/\partial\omega(\uparrow\downarrow) = |\hat{\boldsymbol{\lambda}} \times \boldsymbol{M}_\perp(\boldsymbol{k})|^2. \qquad (3.2)$$

An instrument for doing polarized neutron diffraction experiments is indicated schematically in figure 2. A monochromatic polarized neutron beam is produced by Bragg reflexion of polychromatic unpolarized neutrons from the reactor by a polarizing crystal. The ideal polarizing crystal has $N(\boldsymbol{k}) = M(\boldsymbol{k})$ so that by (3.1) it has zero cross section for neutrons polarized antiparallel to $M(\boldsymbol{k})$ and a large cross section for the reverse polarization. The polarization is maintained in the flight path to the specimen by magnetic guides, producing fields in the polarization direction. At some point in this path a 'spin flipper' is introduced; this may be an r.f. coil tuned to the Larmor frequency, as in the original diffractometer, or one of a number of other devices. The function of the spin flipper is to reverse the polarization when it is 'on' and allow the neutrons to pass unchanged when it is 'off'. In a polarized neutron experiment, the diffractometer and sample are set so that the peak of a Bragg reflexion

[†] 1 Å = 10^{-10} m = 10^{-1} nm.

enters the detector and the ratio between the counting rates for the two polarization states is measured. This ratio, commonly known as the 'flipping ratio' is recorded for each reflexion and these ratios are the 'raw data' of the experiment. From (3.1) and (3.2) it can be shown that the flipping ratio, R, is

$$\frac{|N(\mathbf{k})|^2+|\mathbf{M}(\mathbf{k})|^2+N(\mathbf{k})\,\hat{\boldsymbol{\lambda}}\cdot\mathbf{M}^*(\mathbf{k})+N^*(\mathbf{k})\,\hat{\boldsymbol{\lambda}}\cdot\mathbf{M}(\mathbf{k})}{|N(\mathbf{k})|^2+|\mathbf{M}(\mathbf{k})|^2-N(\mathbf{k})\,\hat{\boldsymbol{\lambda}}\cdot\mathbf{M}^*(\mathbf{k})-N^*(\mathbf{k})\,\hat{\boldsymbol{\lambda}}\cdot\mathbf{M}(\mathbf{k})}. \qquad (3.3)$$

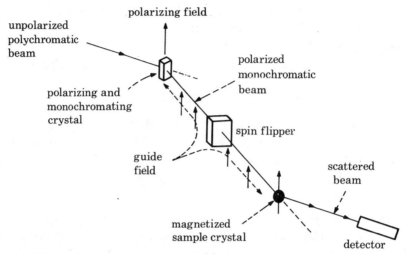

FIGURE 2. Schematic representation of an instrument for polarized neutron diffraction.

In the simple case that both $N(\mathbf{k})$ and $\mathbf{M}_\perp(\mathbf{k})$ are real (centrosymmetric structures) and the polarization is parallel to the magnetization and perpendicular to the scattering vector, the ratio becomes

$$R = \frac{N(k)^2+M(k)^2+2N(k)\,M(k)}{N(k)^2+M(k)^2-2N(k)\,M(k)} = \left(\frac{1+\gamma}{1-\gamma}\right)^2, \qquad (3.4)$$

where
$$\gamma = \frac{M(k)}{N(k)}.$$

In general, R can only be different from unity if there are magnetic and nuclear scattering in the same reflexion so that the cross term in (3.4) is non-zero. This condition is satisfied if any part of the magnetization has the periodicity of the nuclear structure, as in a ferromagnet or a paramagnet in an applied field. It is not satisfied by antiferromagnetic structures in which the magnetic unit cell is some multiple of the chemical one.

The ratio γ can be derived from the flipping ratio, R, by solving the quadratic equation (3.4). The result, which assumes complete polarization and perfect spin reversal, is

$$\gamma = \{R+1\pm\sqrt{(4R)}\}/(R-1). \qquad (3.5)$$

The choice of root depends on whether the absolute magnitude of the magnetic scattering is greater or less than that of the nuclear scattering. There is no uncertainty in the sign of γ, which depends on whether the ratio R is greater than or less than unity. This means that the polarized neutron technique fixes the relative phases of magnetic and nuclear scattering experimentally in centro-symmetric structures. The form of (3.5) shows that γ is not well determined by R when γ is close to 1. In this region, integrated intensity measurements give better accuracy. However, it is one of the major advantages of the polarized technique that integrated intensities are not needed and it is the ratio between the peak counting rates for the reflexion that is measured. In most cases magnetic neutron scattering is much weaker than nuclear scattering and $|\gamma|$ is significantly less than one. For small γ, (3.5) can be simplified as

$$R = 1 + 4\gamma. \tag{3.6}$$

To show the enhanced sensitivity of the polarized technique for measurement of small magnetic scattering, (3.6) can be compared with the fraction of the integrated intensity due to magnetic scattering:

$$\Delta I/I = 1 + \gamma^2. \tag{3.7}$$

The polarized neutron technique can be used to measure extremely small γ since down to 0.1 % accuracy or better there is little except the counting statistics to limit the accuracy with which R can be measured. Measurements of very small γ values do, however, require long counting times and high incident intensity as is shown in table 1. To give some idea of the orders of magnitude involved, the counting rate from a strong reflexion of a simple crystal some tens of cubic millimetres in volume will probably not exceed 10^4/s which implies measurement times per reflexion of 2 min, 3 h and 12 days for the three cases of table 1.

TABLE 1. COUNTING REQUIREMENTS FOR MEASUREMENT OF SMALL γ IN POLARIZED NEUTRON EXPERIMENTS

(For small γ, $\Delta I/I = 1 + \gamma^2$, whereas $R = 1 + 4\gamma$.)

γ	$\Delta I/I$	$R - 1$	count for 5% error in γ
0.01	1×10^{-4}	4×10^{-2}	2×10^6
0.001	1×10^{-6}	4×10^{-3}	2×10^8
0.0001	1×10^{-8}	4×10^{-4}	2×10^{10}

4. A SURVEY OF MEASUREMENTS WITH THE 'CLASSICAL' POLARIZED NEUTRON TECHNIQUE

The 'classical' polarized neutron technique was first described by Nathans *et al.* (1959). Early investigations were confined to simple systems that are ferromagnetic at ambient temperatures, and included detailed studies of the form factors of the metals iron, cobalt and nickel. The development of superconducting magnets and the availability of higher polarized neutron fluxes enabled the technique to be extended to paramagnetic materials and to more complex systems. Many studies have been carried out on the rare earth and actinide elements, their compounds and their alloys, which have yielded new insight into the magnetic states of these materials. In this paper we shall, however, confine our attention to the transition metals and their compounds in which the d electrons not only give rise to magnetism but also play an important rôle in chemical bonding.

Paramagnetic metals of the 3d, 4d and 5d series have been studied at the Oak Ridge National Laboratory in the U.S.A. (Moon *et al.* 1976). Single crystals of the metals were subjected to fields of up to 10 T and the magnetic scattering of the aligned paramagnetic moment was measured. The results provide a very sensitive test of band structure calculations because the magnetic structure factors depend strongly on the eigenfunctions rather than just on the eigenvalues of the electron wave functions.

Relatively few studies of ordered magnetic salts have been made because these usually order antiferromagnetically, and even when mixed magnetic and nuclear reflexions occur, the existence of magnetic domains insensitive to applied fields complicates the experiments. Exceptions include the weak canted ferromagnets typified by $MnCO_3$, in which the spatial distribution of the small ferromagnetic moment has been measured (Brown & Forsyth 1967). Metamagnetic layered halides such as $FeCl_2$ have also been studied, as have some ferromagnetic mixed oxides (Bonnet et al. 1978; Rakhecha & Satya Murthy 1978). The interpretation of the results of these studies has been in terms of ligand field theory of the magnetic ions with the use of parameters obtained from resonance, spectroscopic and bulk magnetic measurements. Differences between the predictions of the models and the measurements demonstrate transfer of magnetic moment to ligand ions and enable simple covalency parameters to be deduced (Hubbard & Marshall 1965).

The availability of higher neutron fluxes, higher magnetic fields and a developing formalism for calculating magnetic structure factors has stimulated the study of more complex magnetic systems in which significant orbital scattering may be present and of more complex chemical systems such as organometallic compounds.

To illustrate the current status of the technique, we shall describe in rather more detail three recent experiments carried out at the Institut Laue–Langevin that exemplify three different areas of application. These are the determination of the paramagnetic form factor of technetium (Radhakrishna & Brown 1979), a study of orbital effects in a ferrimagnetic vanadium salt (Forsyth & Brown 1979) and spin density and bonding in the $[CoCl_4]^{2-}$ ion (Figgis et al. 1979a).

5. THE PARAMAGNETIC FORM FACTOR OF TECHNETIUM

Technetium, which is the 4d analogue of manganese in the 3d and rhenium in the 5d transition metal series, does not occur naturally. It can be obtained from spent nuclear fuel as the radioactive isotope Tc^{99}, a soft β-emitter with a half life of 2×10^5 years. Technetium is a hexagonal close-packed metal; it has a paramagnetic susceptibility which remains essentially constant at 10^{-6} e.m.u./g† over the temperature range 80–1400 K. The form factor of the paramagnetic electrons in technetium is of interest because of the rather strange form factors found for other hexagonal close-packed transition metals such as scandium, titanium and yttrium in which the low angle reflexions do not fall on a smooth curve (Moon et al. 1976).

An applied field of 4.8 T at 100 K gives an aligned paramagnetic moment of 0.94×10^{-3} μ_B per atom of technetium, and the expected γ value at the lowest angle reflexion is about 2×10^{-4}. This shows that measurement of the paramagnetic scattering is at the lower limit of possibility. However, large single crystals have been grown by Kostorz & Mikhailovich (1970) and two were lent to us by the Oak Ridge National Laboratory. These crystals enabled us to achieve counting rates of up to 10^5/s and the rate used was limited only by the detector dead time. Thus with about 28 days measuring time we were able to obtain the flipping ratios of the eight lowest angle reflexions with an accuracy of around one part in 10^{-4}. At this level of precision in flipping ratio measurements, account must be taken of small systematic errors which may invalidate the results. An account of some of these errors is given by Moon et al. (1975). Because of the energy difference between the two neutron polarization states in high fields, and to the

† 1 e.m.u. of magnetic volume susceptibility $\approx 1/4\pi$ of 1 SI unit.

Stern–Gerlach bending of the beam as it traverses the inhomogeneous field regions, the reflexion curves of the two polarization states are not exactly superposed. This may give rise to spurious flipping ratios if the reflexion curve is sharp and the measurement is not made at the exact mean position of the two curves. In the present experiments, the reflexion curves were quite broad (0.25° f.w.h.m.) and an on-line maximization option in the diffractometer control package was used to ensure that the crystal was set to the centre of the reflexion before each measurement. The precision of the flipping ratios was estimated from the degree of reproducibility between repeated measurements of all available equivalent reflexions; in no case was the reproducibility found to be significantly different from that expected from counting statistics.

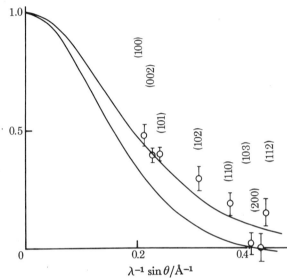

FIGURE 3. The paramagnetic form factor of technetium. The experimental points with their estimated errors are shown. The lower full curve corresponds to the spin-only form factor for Tc 4d electrons. The upper full curve corresponds to the dipole approximation for 70% orbital moment.

Other physical effects that must be taken into account before the paramagnetic structure factors can be obtained from the flipping ratios are extinction, Schwinger scattering and diamagnetism. To determine the degree of extinction a separate experiment with unpolarized neutrons was carried out. The integrated intensities of the reflexions were measured at 100 K in zero field and the results showed extinction to be negligible in the flipping ratios. The magnitude of the Schwinger scattering, which arises from the interaction between the neutrons' spin and the atomic electric field, is in this case of the same order of magnitude as the experimental uncertainty and so has been included. The same is true of the diamagnetic scattering, which is calculated from the second moment of the charge density (Stassis 1970). Once these effects have been removed, the resultant magnetic structure amplitudes are Fourier components of the paramagnetic magnetization density. They are converted to points on the paramagnetic form factor by dividing by the geometric structure factors and the value of the bulk magnetization per cell. The bulk magnetization is also corrected for the diamagnetic moment. The results are shown in figure 3: they fall on a reasonably smooth curve unlike those for scandium, yttrium and titanium. This confirms that the behaviour of these latter elements is due to their early positions in the transition series rather than to their hexagonal structures.

The experimental points lie well above the lower full curve of figure 3 which gives the spin-only form factor for Tc 4d electrons. This is not unexpected since a temperature-independent susceptibility may have an important Van-Vleck orbital component. The experimental data have been fitted to a Tc 4d form factor in the dipole approximation (equation (2.5)) allowing the ratio of spin to orbital moment to vary. The result suggests that 70 % of the moment is orbital, and the resultant form factor is shown in the upper solid curve of figure 3. This model gives a reasonable account of the experimental data; further elaboration must await a good calculation of the technetium band structure.

This study of technetium is an example of the measurement of a very small magnetization in a very simple structure. The results will provide a stringent test of any band structure calculation.

6. ORBITAL EFFECTS IN THE MAGNETIC SCATTERING OF $K_5V_3F_{14}$

Our next example comes from the study of strongly magnetic salts: it shows the dramatic way in which the magnetic scattering can depend on details of the ground state of the magnetic ion. We have already commented that the Mn^{2+} ion exhibits, to a good approximation, the spin-only moment of 5 μ_B and this is also true for the isoelectronic Fe^{3+} ion. Normally $Fe^{2+}(3d^6)$ and $Ni^{2+}(3d^8)$ ions have moments in which there is some 5–10 % orbital contribution, and this proportion increases further for $Co^{2+}(3d^7)$. V^{3+} is an ion in which the 3d shell is less than half full ($3d^2$) and in which we may expect the orbital and spin components, L and S, to be opposed. The Hund's rule ground state is ^3F. The moment associated with the V^{3+} ions in V_2O_3 is 1.2 μ_B (Moon 1970) and this is evidence for a significant orbital contribution which reduces the spin-only moment of 2 μ_B. Balcar et al. (1973) have pointed out that the scattering of V^{3+} in V_2O_3 should be very anisotropic owing to the large orbital contribution. However, a crystallographic phase transition precludes a single crystal study of the magnetization density.

A detailed study is being made of the magnetization of V^{3+} in $K_5V_3F_{14}$, which is one of a series of fluorides having the general formula $A_5^+M_3^{3+}F_{14}^-$, in which A is an alkali metal (Na^+ or K^+) and M is a trivalent transition metal ion (Fe^{3+}, Co^{3+}, V^{3+}). The substances are ferrimagnetic and have chemical structures related to that of chiolite, $Na_5Al_3F_{14}$, which was first determined by Brosset (1938). Chiolite is tetragonal with space group P4/mnc. There are two formula units per unit cell, which has $a = 7.01$ Å and $c = 10.41 \pm 0.01$ Å. The preparation of $K_5V_3F_{14}$ single crystals has been described by Wanklyn et al. (1976) and bulk magnetization measurements show that it becomes ferrimagnetic below 18 K with a saturation magnetization at 0 K corresponding to 2.15 μ_B per formula unit (Cros et al. 1977). The crystals are not damaged by cooling to 4.2 K and two neutron studies have been made at that temperature.

Unpolarized beam diffraction data confirm that $K_5V_3F_{14}$ is isomorphous with chiolite and have enabled its atomic positional and thermal parameters to be determined. A second series of measurements were carried out with the classical polarized beam technique. Flipping ratios of some 1800 reflexions were measured at a wavelength of 0.933 Å. The specimen was in an external field of 1.5 T aligned parallel to the crystal rotation axis, which was [001]. Figure 4 shows that this is the easy direction of magnetization and 0.2 T is sufficient for saturation (Cros et al. 1977). The data were corrected for incomplete beam polarization and averaged over crystallographically equivalent reflexions to yield some 300 independent observations.

The atomic arrangement in $K_5V_3F_{14}$ is illustrated in figure 5. Two crystallographically independent V^{3+} ions are octahedrally coordinated by fluorine atoms and form layers perpendicular to c centred about $z = 0$ and $z = \frac{1}{2}$. The two V(1) atoms at (000) and $(\frac{1}{2}\frac{1}{2}\frac{1}{2})$ have local symmetry $4/mmm$ and have moments that are opposed to those of the four V(2) atoms at $(\frac{1}{2}00)$, $(0\frac{1}{2}0)$, $(\frac{1}{2}0\frac{1}{2})$ and $(0\frac{1}{2}\frac{1}{2})$. The V(2) atoms have orthorhombic local symmetry, with the moment perpendicular to an octahedron edge. The potassium atoms lie between the layers at height $z = \frac{1}{4}$.

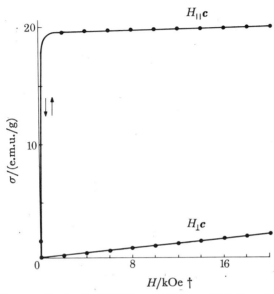

FIGURE 4. The variation of magnetization in $K_5V_3F_{14}$ at 4.2 K as a function of magnetic field H for H parallel and perpendicular to the c axis (Cros *et al.* 1977). † 1 kOe $\approx$ 79.58 kA m^{-1}.

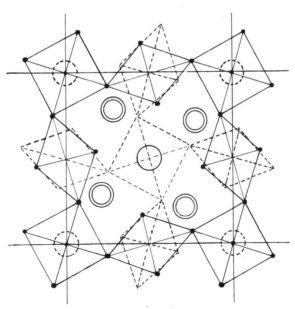

FIGURE 5. The structure of $K_5V_3F_{14}$ projected down [001]. The octahedra of fluorine atoms about the V(1) and V(2) sites are shown, the layer about $z = 0$ in full lines and the layer about $z = \frac{1}{2}$ in broken lines. Potassium atoms, shown as two concentric circles, are at heights $z = \frac{1}{4}$ and $\frac{3}{4}$.

The saturation magnetization of 2.15 μ_B, which arises from the difference in moments $2\mu_{V(2)} - \mu_{V(1)}$, might naïvely be associated with each ion having a moment close to its spin-only value of $2\mu_B$. However, a preliminary analysis of the magnetic scattering amplitudes derived from the two neutron studies showed that both the magnitude and the form factor of the magnetic scattering from each of the independent V sites differ greatly from that to be expected from a spin-only ground state. In particular, the moment on the V(1) site is reduced to some $0.8\mu_B$, whereas that on the V(2) site is about $1.5\mu_B$. Furthermore, the form factor for the V(1) site reaches very large negative values around $\lambda^{-1}\sin\theta$ 0.4 Å^{-1}. However, the form

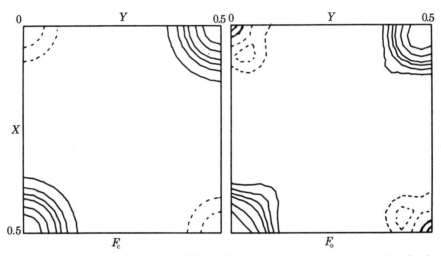

FIGURE 6. The calculated (F_c) and observed (F_o) [001] projections of the magnetization density in $K_5V_3F_{14}$. The calculated density corresponds to a model in which the vanadium ions have free-atom spin-only moments of V(1) = $-0.85\mu_B$ and V(2) = $1.5\mu_B$. The observed magnetization is very aspherical and the moment direction at the centre of the origin V(1) site is in fact oppositely directed to the majority of the magnetization at that site.

vector for the (h, k) even $l = 0$ reflexions, which have structure factors $\mu_{V(1)} + 2\mu_{V(2)}$, approximates more closely to $\langle j_0 \rangle$ for V^{3+}. These results may be partly understood in terms of the dipole approximation, since they would require a reduction in moment from $2\mu_B$ on the V(2) site that is about one-half the reduction for the V(1) site as is observed. The effect of the large orbital contribution on the magnetization density associated with the V(1) site is remarkable, as can be seen in figure 6, which shows its Fourier projection on (001). The projected density close to the atomic position is in fact oppositely directed to that in the outer regions of the ion.

The octahedral crystal field at the V^{3+} ions partially removes the orbital degeneracy of the ^3F state to leave a Γ_4 ground state which is further split by any tetragonal or orthorhombic distortions present at the V(1) and V(2) sites respectively and by the spin–orbit interaction. A calculation of the scattering which would be given by the V(1) atoms if they were in the pure ground state induced by the tetragonal distortion, namely

$$\{\epsilon|-3\rangle + (1-\epsilon^2)^{\frac{1}{2}}|+1\rangle\}|M_s = 1\rangle \quad \text{with} \quad \epsilon = \sqrt{\tfrac{5}{8}}, \tag{6.1}$$

goes some way towards reproducing the sign change in the form factor at $\lambda^{-1}\sin\theta \approx 0.4$ Å^{-1}, but corresponds to $0.5\mu_B$ and not $0.8\mu_B$/vanadium. A comparison between the observed and calculated structure factors for a limited number of reflexions to which only V(1) ions contribute is shown in figure 7. The calculated values correspond to a value for ϵ (equation

(6.1)) modified to give a moment of 0.8 μ_B. The large difference between values corresponding to different l cannot be reproduced by a simple wavefunction of this form. We are now extending the calculation to include the V(2) ions and a Hamiltonian containing the effects of crystal field, magnetic field and spin–orbit coupling. A least squares technique will be used to adjust the amplitudes of a suitable set of basis wavefunctions to minimize the differences between the observed and predicted values of the magnetic scattering amplitudes, and in this respect it resembles the pioneering work of Boucherle (1977) on rare earth intermetallic compounds. It is clear, even at this stage in the interpretation, that the large differences in behaviour of the magnetic scattering from the two very similar crystallographic atomic environments must provide a sensitive probe of their wavefunctions.

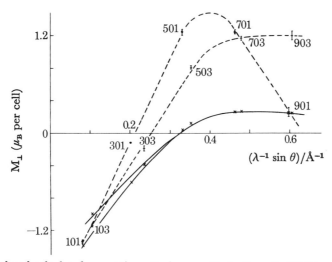

FIGURE 7. Observed and calculated magnetic scattering amplitudes from the V(1) ion in $K_5V_3F_{14}$. In the approximation of spherical distributions, V1 atoms only contribute to the $h0l$ reflexions with l odd. The calculated points (x) correspond to

$$\Psi_{V1} = 0.74|L_z = 3, \ S_z = 1\rangle + 0.67|L_z = -1, \ S_z = 1\rangle$$

which gives a magnetization of 0.8 μ_B/formula unit. The solid lines through the calculated values and the broken lines through the observations serve only to guide the eye.

7. SPIN DENSITIES IN PARAMAGNETIC MOLECULAR CRYSTALS

A direct study of covalence in paramagnetic molecules or ions is offered by the determination of spin density distribution in the crystal via Fourier transformation of magnetic structure factors determined by polarized neutron scattering techniques. A programme of studies of chemically significant compounds is under way (Figgis et al. 1979). However, the limitations attached to quantitative interpretations of Fourier series are well known and Varghese & Mason (1980) have developed the use of least squares multipole electron population analysis of the magnetic data and the relation of such an analysis to conventional orbital descriptions of molecular spin densities.

This approach is illustrated by a discussion of the polarized neutron data obtained from single crystals of Cs_3CoCl_5. Two independent sets of magnetic structure factors have been determined with the magnetization directed respectively along the 'a', ('b') and 'c' crystallographic axes (Figgis et al. 1980a). The crystallography of Cs_3CoCl_5 is well established, both by

X-ray and neutron diffraction methods at 300 and 4.2 K (Figgis *et al.* 1964, 1980*b*): the $(CoCl_4)^{2-}$ ions are embedded in a Cs^+, Cl^- matrix, the coordination geometry of the Co(II) ion being an axially distorted tetrahedron. The Co(II), d^7 ion has essentially a $^4A_2$ ground term; the bulk magnetization estimated from magnetic susceptibility measurements (van Stapele *et al.* 1966) gives a magnetic moment per unit cell of 14.6 μ_B along the crystallographic 'c' axis and 2.56 μ_B along the tetragonally equivalent 'a' and 'b' axes.

A multipole deformation model for the analysis of spin density distributions in transition metal clusters has recently been described (Varghese & Mason 1980). It follows closely the methods established in the context of X-ray determination of molecular charge densities (Stewart 1973; 1976).

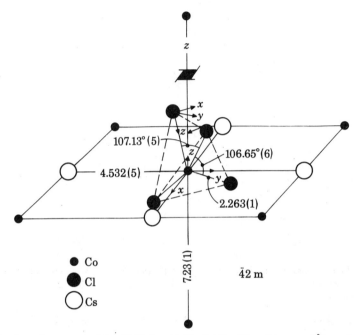

FIGURE 8. The local environment of the Co^{2+} ion in Cs_3CoCl_5. The distances (Å) and the angles are from the structure determination at 4.2 K. The local Cartesian axes used for the multipole analysis lie in the [110] mirror planes, and are related to each other by the point group symmetry at the Co^{2+} site ($\bar{4}2m$).

TABLE 2. TESSERAL HARMONICS (ANGULAR PART OF MULTIPOLE DENSITIES) AND NORMALIZATION CONSTANTS ASSOCIATED WITH THE NON-VANISHING MULTIPOLE POPULATIONS USED IN THE ANALYSIS OF Cs_3CoCl_5 IN TERMS OF THE DIRECTION COSINES (x, y, z) OF THE BRAGG VECTOR RELATIVE TO THE LOCAL CARTESIAN AXES OF THE CENTRES

(The notation used is that of Varghese & Mason (1980).)

l	m	M_l^m	$N_l^m/4\pi$
0	0	1	$1/4\pi$
1	−1	y	0.31831
1	0	z	0.31831
2	−1	$3yz$	0.25
2	0	$\frac{1}{2}(3z^2-1)$	0.4135
2	2	$(3x^2-y^2)$	0.125
3	−2	$30(xyz)$	0.06667
4	0	$\frac{5}{8}(7z^4-6z^2-\frac{3}{5})$	0.55534
4	4	$105(x^4-6y^2+y^4)$	0.004464

The local Cartesian systems centred on the cobalt and chlorine centres are indicated in figure 8; the y and z axes labelling the ligand atoms lie on crystallographic mirror planes. Relative to these Cartesian systems, the non-vanishing multipoles on the cobalt are the scalar M_0^0, the quadrupole M_2^0, the octupole M_3^{-2} and hexadecapoles M_4^0 and M_4^4; on the chlorine atoms we must account for the scalar M_0^0, the dipoles M_1^0 and M_1^{-1} and the quadrupoles M_2^0, M_2^{-1} and M_2^2. Table 2 lists the angular part of the corresponding multipole density function together with normalization factors; the notation is that of Varghese & Mason (1980).

TABLE 3. MULTIPOLE POPULATIONS (ELECTRONS) OF THE SPIN DENSITY DISTRIBUTION IN THE $[\mathrm{CoCl}_4]^{2-}$ ION

(The statistical agreement factors are $R(F) = \Sigma|\Delta F|/\Sigma|F|$ and $\chi = \Sigma\omega\Delta F^2/(N-n)$, where $|\Delta F|$ is the difference between observed and calculated structure factors F, and the sum is over all N observed reflexions including the experimental bulk magnetization F_{000}. R_1 and R_2 refer to the refinements without and with a 4s density term on the cobalt centre, and $\delta\zeta/\zeta$ is the effective contraction or dilation factor of the scalar term M_0^0.)

centre	M_l^m l, m	b-axis data		c-axis data	
		R_1	R_2	R_1	R_2
Co	0, 0	2.39	2.38 (4)	2.53 (4)	2.50 (5)
	2, 0	0.04 (1)	0.04 (1)	−0.15 (3)	−0.27 (4)
	3, −2	0.12 (4)	0.11 (4)	0.13 (1)	0.12 (1)
	4, 0	−0.32 (3)	−0.30 (3)	−0.26 (5)	−0.17 (5)
	4, 4	−0.36 (5)	−0.40 (6)	−0.205 (7)	−0.219 (7)
	$\delta\zeta/\zeta$	0.144 (2)	0.153 (2)	0.160 (3)	0.200 (4)
	0, 0 (4s)		0.10 (6)		0.40 (4)
Cl	0, 0	0.18 (1)	0.17 (2)	0.066 (9)	0.35 (8)
	1, 0	−0.007 (5)	−0.006 (6)	0.007 (2)	−0.018 (2)
	1, −1	−0.003 (5)	−0.003 (5)	0.011 (3)	−0.001 (3)
	2, 0	−0.004 (6)	−0.005 (6)	0.002 (3)	−0.006 (3)
	2, −1	−0.02 (1)	−0.02 (1)	0.036 (8)	0.010 (9)
	2, 2	−0.004 (6)	−0.004 (6)	−0.002 (2)	0.006 (2)
	$\delta\zeta/\zeta$	−0.24 (2)	−0.25 (2)	−0.17 (2)	+0.25 (2)
	$R(F)$	0.051	0.051	0.037	0.029
	χ	1.269	1.260	2.370	2.049
	number of reflexions, N	111	111	78	78
	number of variables, n	13	14	13	14
	F_{000} cal./μ_B	2.561	2.561	12.98	14.15

TABLE 4. ORBITAL POPULATIONS (ELECTRONS) OF THE SPIN DENSITY DISTRIBUTION IN THE $[\mathrm{CoCl}_4]^{2-}$ ION IN $\mathrm{Cs}_3\mathrm{CoCl}_5$, OBTAINED FROM THE MULTIPLE REFINEMENTS LISTED IN TABLE 3 BY USING THE TRANSFORMATION EQUATIONS OF VARGHESE & MASON (1980)

centre	orbital	b-axis data		c-axis data	
		R_1	R_2	R_1	R_2
Co	d_{z^2}	0.07 (6)	0.10 (5)	−0.02 (5)	−0.02 (5)
	$d_{x^2-y^2}$	−0.09 (8)	−0.26 (9)	0.28 (6)	0.40 (6)
	d_{xy}	0.93 (8)	0.99 (9)	0.92 (6)	1.08 (6)
	d_{xy}, d_{yz}	0.79 (3)	0.78 (3)	0.67 (6)	0.52 (6)
Cl	p_x	0.06 (1)	0.06 (1)	0.02 (1)	0.01 (1)
	p_y	0.06 (1)	0.05 (1)	0.03 (1)	0.01 (1)
	p_z	0.06 (1)	0.06 (1)	0.03 (1)	0.01 (1)

Radial wave functions are the self-consistent values for Co^{2+} (3d) and Cl (3p) given by Clementi & Roetti (1974). In the least squares analysis, the effective nuclear charge or the nephalauxetic effect ($\delta\zeta/\zeta$) was refined by taking a first-order Taylor expansion of the radial function. A scalar population $M_0^0(4s)$, with a self-consistent 4s radial function, was included in the final refinement.

Table 3 lists the results of the multipole population analysis while the interpretation of these populations, in terms of 'd' electron populations at the cobalt ion and '3p' electron density at the chlorines, is shown in table 4. If the spin density arises purely from one-centred d- and p-density functions on the metal and ligand centres respectively, then the multipoles M_3^{-2} (cobalt) and M_1^0, M_1^{-1} and M_2^{-2} (chlorine) should be negligible compared with others. The results are generally consistent with this expectation except that the dipole and quadrupole terms emerging from the 'c' axis refinement are statistically significant and represent the projection of two-centre overlap density on to the ligand centres.

Both data sets confirm the t_2^3 configuration at the metal with an essentially spherically symmetric distribution of density at the ligands. Differing covalence implied from the 'b' axis data refinement reflects, we believe, systematic errors in the analysis: there is a poorer sampling of ligand spin owing to the lower accuracy of scattering data relating to covalence (the 'c' axis data contain (hkl) reflexions with l odd, which receive contributions only from the ligand atoms and any acentric density based on the cobalt ion).

There is a 15% contraction of the radial wave function of cobalt vis-à-vis the free ion value, the contraction results from both the transfer of spin to the ligands and the effect of the angular momentum contribution to the magnetization density. Within the dipole approximation (Marshall & Lovesey 1971), one expects a small modification to the radial part of the scalar density terms dependent on g: this has the effect of contracting the scalar density by approximately one-half that observed. It may be possible to infer that the orbital contribution to both data sets is similar on account of the similar g values and the magnitude of the orbital contractions in the two cases. This is a tentative conclusion that needs further study with more complete data and a clearer understanding of how our fitting approach may disguise an orbital contribution. Indeed, the apparently significant difference in population of the metal 3d(t) orbitals, which emerges from the 'c' axis refinement, may be a direct indication of an inadequate modelling of orbital contributions.

Inclusion of a 4s density function in the analysis gives a markedly improved calculated bulk magnetization for the 'c' axis data with a population of 0.40 (4) μ_B. This inclusion may sample the charge in the overlap region, a view which is supported by some *ab initio* calculations on the $(CoCl_4)^{2-}$ ion (Hillier *et al.* 1976). But we have not yet confirmed the uniqueness of an improved analysis with a 4s metal contribution compared with that resulting from including a δ-function on the metal.

The general conclusions are straightforward: the 'c' axis data provide an accurate description of ligand and overlap densities while the 'b' axis data give precise spin densities at the cobalt. The cobalt has the t_2^3 configuration, an orbital contraction of 8% and a net spin of 2.34 (4) μ_B; 0.07 (1) μ_B are transferred to the chlorine centre, the ligand density being essentially spherically symmetric. A plausible remaining interpretation is that the remaining spin (0.40 (4) μ_B is distributed in a diffuse metal 4s orbital.

REFERENCES (Brown et al.)

Balcar, E., Lovesey, S. W. & Wedgewood, F. A. 1973 *J. Phys.* C **6**, 3746.
Bonnet, M., Delapalme, A., Becker, P. & Fuess, H. 1978 *J. Magn. magn. Mater.* **1**, 23.
Boucherle, J.-X. 1977 Thèse pour Docteur ès-sciences physiques, Université de Grenoble.
Brosset, C. 1938 *Z. anorg. allg. Chem.* **238**, 201.
Brown, P. J. & Forsyth, J. B. 1967 *Proc. phys. Soc.* **92**, 125.
Clementi, E. & Roetti, C. 1974 *Atom. data nucl. Data Tables* **14**, 177.
Cros, C., Dance, J.-M., Grenier, J.-C., Wanklyn, B. M. & Gerrard, B. J. 1977 *Mater. Res. Bull.* **12**, 415.
Figgis, B. N., Gerloch, M. & Mason, R. 1964 *Acta crystallogr.* **17**, 506.
Figgis, B. N., Mason, R., Reynolds, P. A., Smith, A. R. P., Varghese, J. N. & Williams, G. A. 1980a *J. chem. Soc. Dalton Trans.* (In the press.)
Figgis, B. N., Mason, R., Smith, A. R. P. & Williams, G. A. 1980b *Acta crystallogr.* (In the press.)
Figgis, B. N., Mason, R., Smith, A. R. P. & Williams, G. A. 1979 *J. Am. chem. Soc.* **101**, 3673.
Forsyth, J. B. & Brown, P. J. 1979 (In preparation.)
Hillier, I. H., Kendrick, J., Mabbs, F. E. & Garner, C. F. 1976 *J. Am. chem. Soc.* **98**, 395.
Hubbard, J. & Marshall, W. C. 1965 *Proc. phys. Soc.* **86**, 561.
Kostorz, G. & Mikhailovich, S. 1970 In *Proc. 12th Int. Conf. on Low Temp. Phys.*, p. 341. Academic Press of Japan.
Marshall, W. & Lovesey, S. W. 1971 In *Theory of thermal neutron scattering*, p. 156. Oxford University Press.
Moon, R. M. 1970 *Phys. Rev. Lett.* **25**, 527.
Moon, R. M., Koehler, W. C. & Cable, J. W. 1976 In *Proceedings of the Conference on Neutron Scattering*, Gatlinberg, Tennessee (CONF-760 601-P2) (ed. R. M. Moon), p. 577. National Technical Information Service, U.S. Department of Commerce, Springfield, Virginia.
Moon, R. M., Koehler, W. C. & Shull, C. G. 1975 *Nucl. Instrum. Meth.* **129**, 515.
Nathans, R., Shull, C. G., Shirane, G. & Andresen, A. 1959 *J. Phys. Chem. Solids* **10**, 138.
Radhakrishna, P. & Brown, P. J. 1980 *J. Phys.* F. **10**, 489.
Rakhecha, V. R. & Satya Murthy, S. N. 1978 *J. Phys.* C **11**, 4389.
van Stapele, R. P., Beljers, H. G., Bongers, P. F. & Zijlstra, H. 1966 *J. chem. Phys.* **44**, 3719.
Stassis, C. 1970 *Phys. Rev. Lett.* **24**, 1415.
Stewart, R. F. 1973 *J. chem. Phys.* **58**, 1668.
Stewart, R. F. 1976 *Acta crystallogr.* A **32**, 565.
Varghese, J. N. & Mason, R. 1980 *Proc. R. Soc. Lond.* A **372**, 1.
Wanklyn, B. M., Gerrard, B. J., Wondre, F. & Davidson, W. 1976 *J. Cryst. Growth* **33**, 165.

Precise location of hydrogen atoms in complicated structures by diffraction of polarized neutrons from dynamically polarized nuclei

By M. Leslie[†], G. T. Jenkin[†], J. B. Hayter[†], J. W. White[†], S. Cox[‡] and G. Warner[‡]

[†] *Institut Max von Laue – Paul Langevin, Avenue des Martyrs, 38042 Grenoble cedex, France*
[‡] *Science Research Council, Rutherford Laboratory, Chilton, Didcot, Oxon. OX11 0QX, U.K.*

The first results are reported on the precise location of hydrogen atoms in a complicated crystal structure by using diffraction of polarized neutrons from dynamically polarized protons in the sample. Two methodologies exploiting the spin-dependent scattering power of protons are briefly illustrated.

Introduction

A method equivalent to isomorphous substitution techniques in X-ray diffraction can be used for locating hydrogen atom positions in complex crystals as a result of the strong spin-dependent scattering length of the hydrogen atom (Hayter *et al.* 1974). This technique has now been developed to determine precisely the positions of the hydrogen atoms in lanthanum magnesium nitrate hydrate, a crystal that was chosen to display the methodological aspects of this technique.

The crystal of lanthanum magnesium nitrate hydrate ($La_2Mg_3(NO_3)_{12}\cdot 24H_2O$) was grown from extremely pure solutions of the chemicals and of water with a 1% substitution of $^{142}Nd^{3+}$. This gave a high-grade crystal (typical mosaic block spreads of the order of fractions of a minute of arc) containing about 0.1% of neodymium atoms randomly occupying the lanthanum sites of the crystal. In a conventional dynamic nuclear polarization apparatus (Jenkin *et al.* 1976) this gave enhancements of the proton polarization of factors of 600 with the use of the solid state effect (Abragam & Goldman 1978), that is to say, dynamic nuclear polarization of greater than 90%. This is an indication of the high quality of the crystal and its freedom from other impurity ions.

So the crystal represents an ideal subject on which to test the methodological aspects of the use of dynamic polarization to detect hydrogen positions because the polarization can be varied over a wide range to allow this to be used as a variable in the crystallographic measurements, and because the effects on the polarization and on the flip ratio measurements due to extinction, multiple scattering, etc., are likely to be worst for this crystal than for most crystals on which it might subsequently be wished to apply this technique. In this case, if it can be shown that the results are valuable and interpretable, then it can be expected that under the correct conditions the technique of using nuclear polarization and polarized neutron scattering may be of more general applicability. In a more detailed report, the description of the way in which multiple scattering, extinction and other crystallographic effects have been treated to produce the results shown in this brief report will be given.

For the experiment a crystal of the order of 1 mm³ in volume was mounted in an all-aluminium multi-mode microwave cavity operating at approximately 4 mm wavelength. The cavity and crystal were placed in a superconducting coil magnet and could be refrigerated to temperatures as low as 0.7 K. Dynamic nuclear polarization was obtained by adjusting the field and frequency conditions for saturation of the forbidden electron spin resonances in the dipolar electron nuclear coupled system, and the polarization was monitored by using nuclear magnetic resonance and an integrator system. Continuous digital readout of polarization and field conditions were maintained throughout the experiment and were taken into account in the assessment of data. The whole experiment was mounted in a cryostat such that a polarized neutron beam could be scattered from the sample (Hayter *et al.* 1974) and reflexions from the dynamically polarized crystal at two polarizations were collected, both in the zero layer (xz) plane and in the first layer. In order to have a check of the correct crystal structure for lanthanum magnesium nitrate, which should be available from the refinement procedure, the full neutron diffraction determination of the hydrogen positions in the same crystal was made by using conventional neutron scattering techniques (Anderson *et al.* 1977). For the actual experiment, flip ratios at the maxima in the structure factor were recorded for the incident neutron spin up or down with respect to the polarization direction of the polarized nuclei. Nuclear polarization of 20% was used, although some runs were made at 40% nuclear polarization. In all, a data set comprising some 500 independent flip ratio measurements on 500 independent reflexions was measured.

Results

For a preliminary inspection of the data, a difference Fourier synthesis was made by using the spin flip data. This was achieved by taking the known 4 K structure determined by neutron diffraction as a basis and using the known spin dependent scattering length for hydrogen $b^+ = 1.08 \times 10^{-14} M$, $b^- = -4.74 \times 10^{-14} M$. The expected neutron spin up structure factors were calculated from this basic information. The neutron spin down structure factors were then calculated by using the measured flip ratios in the experiment and a hydrogen nuclear scattering density Fourier pattern for the xz plane calculated. This is shown in figure 1a, and for comparison (figure 1b) the similarly calculated Fourier pattern generated from the known low-temperature structure without including the neutron spin flip measurements.

It is immediately apparent that the resolution on the data with the measured spin flip information included is lower than that generated from conventional neutron diffraction measurements alone. This is the first indication of the effects of polarization distribution in the spin polarized crystal and will be discussed in more detail later. Nevertheless, it is apparent that, by using the relatively small number of reflexions taken in the spin flip data, all of the hydrogen atoms seen in the conventional structure can, in principle, also be seen by the use of the spin slip information.

The approach next adopted was more akin to that which would be used for solving an unknown structure, but given some basic information from X-ray diffraction. It is our belief that, for using this technique, a good point of departure at this stage should be a good X-ray structure which would find atoms such as oxygen and nitrogen in enzyme crystals, for example. It is not worthwhile to use neutron scattering to determine these positions since the X-ray method is faster and cheaper.

The structure $Ce_2Mg_3(NO_3)_{12}.24H_2O$ was therefore taken as a point of departure, the structure determined from X-rays by Zalkin *et al.* (1963). By comparison with our own low-temperature diffraction work, this structure is not completely correct for $La_2Mg_3(NO_3)_{12}.24H_2O$, especially in the location of hydrogen atoms, which they did by a difference Fourier synthesis on the X-ray patterns. Nevertheless, the oxygen and nitrogen, lanthanum, magnesium and other atomic positions except hydrogen are a reasonable basis for departure in our second approach. Here, the hydrogen positions were guessed on the basis of chemical knowledge of OH bond lengths and of hydrogen bond lengths. It was reckoned that these guesses might be good to a few tenths of an ångström† in all cases.

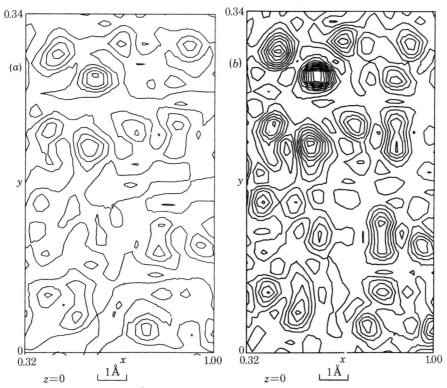

FIGURE 1. (a) The difference Fourier synthesis locating the hydrogen atoms in the xy projection of the structure of lanthanum magnesium nitrate hydrate at 1.2 K. (b) The difference Fourier map of the hydrogen positions obtained from classical neutron diffraction data.

The guessed hydrogen positions were then used for a least-squares refinement of the flip ratios as measured. Data from both the zero layer and the first layer were taken, although those data from the first layer were less reliable than those from the zero layer. After a small number of iterations the structure refined to a good minimum ('R factor' 12.5%). Table 1 compares the observed flip ratios, r, for several reflexions with those calculated from the structure at $R = 12.5\%$. The hydrogen positions had moved to those as determined by the conventional neutron scattering technique. This shows the success of the method and it therefore remains only to demonstrate the degree to which the limited amount of data available on the flip ratios (due to lack of extensive measuring time) can be used to pull in the hydrogen positions from a distance.

To test this point, single hydrogen atoms were moved to various distances from their equilibrium positions along various directions. In all cases, when a single hydrogen atom was moved the refinement procedure produced the correct structure after of the order of 5–15 iterations.

† 1 ångström (Å) = 10^{-10} m = 10^{-1} nm.

A plot of the refinement factor (on the spin flips) as a function of the displacement of the H_{11} hydrogen atom from its equilibrium position in the z and the y and the x directions is shown in figure 2. It is immediately clear that the hypersurface for the refinement is complicated. This may well be a consequence of the limited data set being used. Figure 3 shows a mapping of this hypersurface, giving the true deep minimum of the correct structural position for this atom and indications of subsidiary minima, which undoubtedly arise because of the lack of high-angle reflexions. (For the particular spectrometer used in these measurements, D3 at the Institut Laue–Langevin, Grenoble, high-angle reflexions were difficult to take because of the small take-off angle of the instrument which makes the resolution rather poor at high angles.)

TABLE 1

reflexion			r_{obs}	r_{calc}
0	0	33	0.772	0.594
0	0	30	1.542	1.598
0	0	27	15.552	5.380
0	0	24	4.919	4.736
0	0	21	0.248	0.263
0	0	18	0.842	0.777
0	0	15	0.440	0.752
0	0	12	3.278	3.232
0	0	9	0.441	0.529
0	0	6	0.421	0.439
0	0	3	0.336	0.324
1	1	-6	1.586	1.820
1	1	-12	0.629	0.668
1	1	-15	5.512	7.383
1	1	-18	1.141	1.174
1	1	-21	1.405	1.188
1	1	-24	0.194	0.231
1	1	-27	0.741	1.347
1	1	-30	1.029	2.994
1	1	-33	2.283	2.500
1	1	-36	1.246	1.112

If more than one atom is moved at a time, the structure refines to the correct value up to displacements of the order of 0.5 Å from the equilibrium position, so long as all atoms are moved in the same direction. Preliminary indications are that, if atoms are moved in different directions, the structure does not refine as well as this but may find false minima.

Polarization distribution

There is good evidence from other work that the polarization in a dynamically polarized crystal is not uniform (Abragam *et al.* 1972). In our experiment the flip ratio may be written

$$r = \left(\frac{A+Bp}{A-Bp}\right)^2,$$

where A is the conventional structure factor, B the structure factor of hydrogen atoms alone with $b_H = \frac{1}{4}(b_+ - b_-)$, and p the proton polarization.

Clearly, for certain reflexions it may be possible to have $A = Bp$, and hence r should rise to a very large value. In practice, r is limited by such factors as the neutron beam polarization and flipper efficiency, which can be measured by replacing the sample with a FeCo crystal. Multiple scattering will also limit the maximum (or minimum) flip ratio that can be observed.

We have measured r as a function of polarization for several reflexions at two different temperatures. The data for the $(0,0,12)$ reflexion are shown in figure 4. This shows that there is an additional temperature-dependent effect. We attribute this to the distribution of polarization in the crystal. The distribution becomes less homogeneous at the higher temperature and hence the minimum value of r increases.

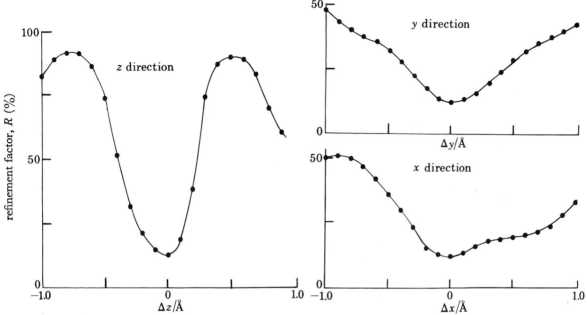

FIGURE 2. Refinement factor (R) obtained from a least-squares refinement of the position of the proton H_{11} against the measured flip ratios in the basal and the first layer plane of the diffraction pattern of lanthanum magnesium nitrate hydrate. The three curves are appropriate for displacements of the proton in the z, x and y directions. Similar behaviour is found for displacements of other single protons in the unit cell.

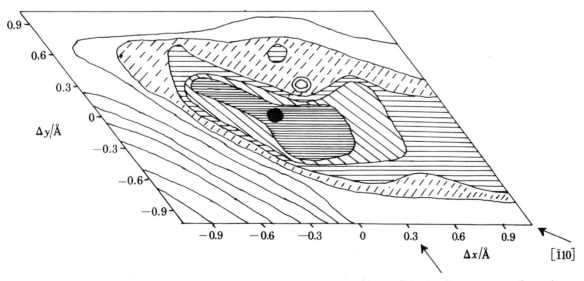

FIGURE 3. Contour diagram showing the true minimum in the xy plane of the lanthanum magnesium nitrate hydrate structure obtained by a least squares refinement against the spin slip ratios.

Further evidence for the polarization distribution came from measurements made as the crystal is polarized then depolarized. Figure 5 shows a hysteresis in the measured intensities. The minimum for the spin down case is greater when the crystal is polarizing. This is when the inhomogeneities are expected to be greatest. Finally, for our measurements at constant polarization we assume that a polarization distribution is still present. We define the distribution by a parameter α, such that

$$\bar{p}^2 = (1+\alpha)\,\bar{p}^2,$$

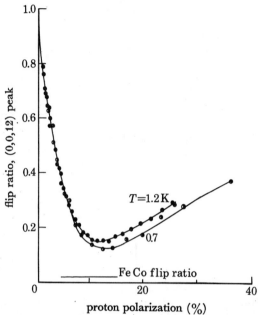

FIGURE 4. Flip ratio of the (0,0,12) reflexion as a function of proton polarization at two temperatures.

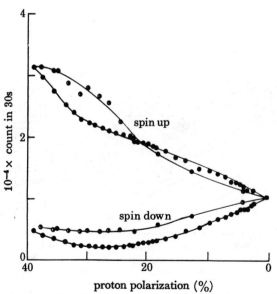

FIGURE 5. Spin up and spin down intensity of the (3,0,12) reflexion at 1.2 K during polarization and depolarization. ◐, Crystal polarizing; ●, crystal depolarizing. All counts corrected for background.

where $\bar{p}$ is the average proton polarization and $\bar{p}^2$ the average of the square of the proton polarization.

The value of α is fitted from our measured flip ratios. Typical values are $\alpha = 0.25$ for $\bar{p} = 40\%$.

Preliminary small-angle neutron scattering experiments have been made to detect the direct scattering of polarized neutrons from the polarization distribution in the crystal.

Discussion

We conclude from these measurements that the use of nuclear polarization as a method of labelling protons in a complicated unit cell may have advantages for determining the proton structure. This would typically be of interest, for example, in determining the proton configuration around the active site in an enzyme. A relatively limited set of data plus the X-ray coordinates of heavier atoms can be used to obtain a refinement of good precision (proton location typically to within 0.05 Å, of the structure determined by classical neutron diffraction measurements at low temperatures). An interesting physical phenomenon discovered is the variation of nuclear spin polarization throughout the crystal. Inevitably this must give rise to a broad background to the diffraction pattern which would, in general, be the Fourier transform of the spin distribution. This broad background in our present measurements leads to a loss of resolution which could, in principle, be recovered by making a model of the spin distribution and refining again the distribution parameter which may be of intrinsic interest itself in a discussion of the physics of highly polarized samples.

References (Leslie et al.)

Abragam, A., Bacchella, G. L., Glattli, H., Meriel, P., Piesvaux, J., Pinot, M. & Roubeau, P. 1972 In *Proceedings of the* 17th *Congress Ampere*, Turku, Finland. Amsterdam: North-Holland.

Abragam, A. & Goldman, M. 1978 *Rep. Prog. Phys.* **41**, 395.

Anderson, M. R., Jenkin, G. T. & White, J. W. 1977 *Acta crystallogr.* B **33**, 3933–3936.

Hayter, J. B., Jenkin, G. T. & White, J. W. 1974 *Phys. Rev. Lett.* **33**, 696.

Jenkin, G. T., Seyfert, P., Feltin, D., Ragazzoni, J. L., Roubeau, P., Banks, P., Cox, S. & Harris, N. 1976 I.L.L. Internal Report no. 76J237T.

Zalkin, A., Forester, J. D. & Templeton, D. H. 1963 *J. chem. Phys.* **39**, 2881.

Neutron diffraction studies of proteins

By G. A. Bentley and S. A. Mason
Institut Max von Laue – Paul Langevin,
156X Centre de Tri, 38042 Grenoble cedex, France

Neutrons interact differently with protein crystals than do X-rays. Not only do hydrogen or deuterium atoms diffract neutrons relatively more strongly, but in addition protein crystals suffer no radiation damage in a neutron beam. These and other differences are being exploited for a few selected proteins, at three reactors.

At the Institut Laue–Langevin, Grenoble, the crystal structure of the triclinic form of hen egg-white lysozyme is being refined at high resolution (d-spacings down to 1.4 Å (0.14 nm)), from neutron diffraction measurements on a partly deuterated native crystal of volume 20 mm^3. Results at the present stage of refinement are discussed. Prospects for neutron protein crystallography are examined in the light of progress in X-ray protein crystallography and neutron detector technology.

Introduction

The many successes of X-ray crystallographic studies of large molecules such as proteins, nucleic acids and viruses are too well known to need exposition here. It is sufficient to recall the pioneering work on myoglobin, haemoglobin, lysozyme or insulin, culminating in the 1960s and 1970s in high-resolution structural models of the molecules and their surroundings in the crystals. It seems particularly appropriate to refer specifically to a Discussion Meeting on the structure and function of lysozyme, organized on behalf of the Royal Society 14 years ago by M. F. Perutz, since lysozyme was the first enzyme for which the crystal structure was determined at high resolution (Phillips 1966; Blake *et al.* 1967 *a, b*). Imoto *et al.* (1972) have reviewed in detail the structure and properties of lysozymes.

Further X-ray crystallographic work is in progress on the lysozyme structures. High-resolution refinements of the structures of tetragonal hen egg-white lysozyme and human lysozyme are near completion. A wealth of information should soon be available about the water structure around the lysozyme molecules, and about the flexibility of different parts of the protein molecules (Artymiuk *et al.* 1979). Of most direct relevance to the mechanism of action of lysozyme is the planned study of substrate complexes of a tortoise lysozyme (C. C. F. Blake, personal communication), since this lysozyme is apparently unique in crystallizing with the entire active cleft region exposed to solvent. Thus for the first time one should be able to study, by crystallogaphic methods, several steps in the pathway of reaction of lysozyme with a complete substrate analogue. Such an experiment would provide the best possible data for comparison with theoretical studies on the energetics of enzyme reactions.

The present paper considers progress in our study of native triclinic lysozyme by neutron diffraction, and the potential of this technique in protein crystallography. We have deliberately kept the discussion of our own work as brief as possible since the refinement is still in progress and has been considered elsewhere (Bentley *et al.* 1979).

Neutron diffraction and lysozyme

That protein crystals may usefully be studied by neutron diffraction was first shown for myoglobin (Schoenborn 1969; Norvell et al. 1975). Whereas X-ray crystallography is most convenient for definition of the protein non-hydrogen atoms, neutron diffraction can give much more information about the positions and orientations of water molecules and salt ions, as well as confirming the positions of hydrogen atoms and showing which hydrogens are exchangeable with deuterium.

Table 1. Some parameters of triclinic lysozyme

hen egg-white lysozyme, 129 residues
crystallized from 2% $NaNO_3$ 1% lysozyme, 0.025 M acetate buffer, pH 4.6, by seeding
triclinic, space-group P1, $Z = 1$
unit-cell volume 26 500 Å^3, molec. mass 20 200
one lysozyme molecule (molec. mass 14 600), 300 H_2O (27% water)
crystal volume, up to 20 mm^3
partial deuteration at pD 4.6, by soaking

Table 2. Neutron measurements on lysozyme

instrument D8, 4-circle neutron diffractometer
single detector, high-pressure ^{3}He
wavelength 1.68 Å, flux 2×10^8 n cm^{-2} s^{-1} ($\lambda/2$ component about 4.8% at detector)
31-step omega scans, small detector aperture

deuterated crystal
 $d > 1.7$ Å, 1–3 min per reflexion
 1.7 Å $\geqslant d \geqslant$ 1.4 Å, 3–5 min per reflexion
 14735 of 17768 unique reflexions have $I > 2\sigma(I)$
 crystal volume 20 mm^3, $T = 23$ °C

native crystal (i.e. not deuterated)
 $d > 1.8$ Å, 1–3 min per reflexion
 crystal volume 20 mm^3, $T = 5$ °C

Table 3. Neutron and X-ray scattering factors (femtometres) (Bacon 1975)

element	neutron	X-ray ($\sin \theta = 0$)
H	−3.7	2.8
D	6.7	2.8
C	6.7	16.9
N	9.4	19.7
O	5.8	22.5
Na	3.6	30.9
S	2.8	45.0

At the Institut Laue–Langevin reactor in Grenoble, we have made neutron diffraction measurements on triclinic lysozyme, for which we give some pertinent parameters in tables 1 and 2. Further details are available elsewhere (Bentley et al. 1979). It is worth noting that for the partly deuterated lysozyme crystals, only about 3½ months of neutron time were needed to record, with moderately good statistical precision, most of the reflexion data with d-spacings larger than 1.4 Å†.

Thus triclinic lysozyme is ideal for high-resolution neutron work, although many other factors combined to dictate the choice of this system (Bentley et al. 1979); with it we should be able to show convincingly how much new information can in practice come from neutron diffraction.

† 1 Å = 10^{-10} m = 10^{-1} nm.

For proteins, the fundamental advantage of neutrons over X-rays is that the neutron scattering length of hydrogen is about 60 % of those of oxygen and carbon, and that of deuterium is about the same as that of carbon (see table 3). Thus hydrogen, and especially deuterium, are relatively much stronger scatterers of neutrons than of X-rays. For this reason, hydrogen and deuterium atoms are visible in properly phased neutron Fourier maps, provided their temperature factors are not too high.

Further, neutrons do not damage protein crystals, so that only one crystal is needed for the entire experiment. This, and the fact that measurements at low temperatures are routine in neutron diffraction, could be of great advantage for experiments such as that on tortoise lysozyme referred to earlier.

Neutron refinement of the triclinic lysozyme structure

Our neutron refinement based at present only on the data from deuterated crystals, started from a structural model determined from X-ray diffraction data by Jensen and coworkers (Kurachi *et al*. 1976; L. H. Jensen, personal communication). After a first attempt to introduce most of the hydrogen atoms at calculated positions, we tried the more cautious approach of starting from the non-hydrogen protein atoms as found in the X-ray work, and gradually including hydrogen or deuterium atoms, and also solvent atoms, in the model. The 1730 atoms placed so far include 165 solvent atoms and 623 hydrogen or deuterium atoms. The agreement factor after idealization of geometry is

$$R = \Sigma(|F_o| - |F_c|)/\Sigma |F_o| = 0.28$$

for data with d-spacings greater than 1.5 Å. Here $|F_o|$ and $|F_c|$ are the observed and calculated neutron structure amplitudes.

Our model is far from complete, and in particular we have made no deliberate attempt to fit complete D_2O molecules to the solvent density. Such fitting will certainly be possible for some of the solvent regions as found also for myoglobin (Schoenborn & Hanson 1979) and trypsin (Kossiakoff & Spencer 1979).

As has been pointed out by Finney (1979) in a comprehensive review of water structure and proteins, it is easy to over-interpret apparent electron density in the solvent regions around protein molecules in crystals. On the other hand, there is no fundamental reason why, in favourable cases such as triclinic lysozyme, an objective description in conventional crystallographic terms (positional and thermal parameters, and occupation factors to allow for disorder and in the neutron case for exchange with deuterium) should not be given of a large proportion of the solvent regions. Serious attempts are now being made to find criteria for the reality of these water molecules (see, for example, Watenpaugh *et al*. 1978) and neutron diffraction can make a unique contribution (Schoenborn & Hanson 1979).

For triclinic lysozyme we expect that direct comparisons of the nuclear scattering density in the deuterated crystal and in the non-deuterated crystal (for which the 1.8 Å resolution data are now being analysed), with the electron density from the X-ray work of Jensen and coworkers, will greatly help our interpretation of the solvent regions. Another approach to examining apparent water structure is to compare the relative positions of solvent atoms in different crystal forms of the same or similar proteins. A beginning has been made for lysozyme in the tetragonal and triclinic crystals (Moult *et al.* 1976), but this approach needs to be followed up by comparing the many high-resolution maps of various similar lysozymes now available.

In view of the current interest in the meaning of the apparent temperature factors in lysozyme (Artymiuk *et al.* 1979), we note in passing that in the regions of our triclinic lysozyme model that have refined well there is reasonable qualitative agreement with the thermal parameters in tetragonal hen egg-white lysozyme and in human lysozyme (Artymiuk *et al.* 1979). Thus, for example residues 13–23, 47 and 70–75 have relatively high temperature factors. On average, the temperature factors in the triclinic crystals are somewhat lower (overall B about 9.5 Å^2) than those in the other lysozymes, and further the packing of molecules in the triclinic case is quite different.

Good correspondence of the temperature factor curves, if confirmed at a later stage in our refinement, would therefore add further weight to the hypothesis that molecular flexibility is relevant to the biological role of proteins.

THE COMPLEMENTARITY OF X-RAY AND NEUTRON DIFFRACTION

Many man-years of effort have gone into various approaches to data collection and crystallographic analysis and refinement in the neutron diffraction studies of myoglobin. The calculations were based on the original X-ray model established over 10 years ago when refinement techniques were relatively unsophisticated. It is certain that much tedious manual interpretation of Fourier maps could in the future be avoided by using the better quality X-ray models now available.

For the neutron work on ribonuclease A (Wlodawer 1979) it has been found necessary to record new X-ray data and further refine the X-ray structure to obtain an adequate starting model for the neutron refinement.

New work on trypsin now in progress at Brookhaven appears to show that with a highly refined X-ray model, even neutron data of modest resolution may help to answer some important biological questions. The analysis, based so far on neutron data to 2.2 Å resolution, has already led to determination of the correct orientations of the amide side chains that are well ordered, as well as showing unambiguously the position of an important proton in the active site (Kossiakoff & Spencer 1979; A. A. Kossiakoff, personal communication).

Taken together with our own experience in refining the lysozyme structure, all of the above experiments suggest that to expedite neutron refinements one should use to the maximum the X-ray diffraction data, which are both cheaper and inherently simpler to interpret since the hydrogen atoms are of minor importance in the refinement. A good starting model for the refinement is even more important in work with neutron diffraction since it is necessary to begin with well placed hydrogen atoms; the ratio of intensity observations to parameters refined is thus reduced by about a half. For example, considering the protein alone, this ratio is 4.5 for the X-ray refinement and 2.3 for the neutron refinement with data to 1.4 Å *d*-spacing.

BIOLOGICAL MOLECULES AT HIGH RESOLUTION

In low to medium resolution studies of proteins (say $d > 2$ Å) it is quite normal to omit hydrogen atoms from the protein model in work with X-ray diffraction, but for neutrons this is unsatisfactory since hydrogen and deuterium atoms make up more than half of the unit cell contents. At medium to high resolution ($d \approx 1.5$ Å), it is advisable to include protein hydrogen atoms in the X-ray model, at predicted positions; and it is imperative in the neutron model if the neutron intensity data are to be explained.

At very high resolution, $d < 1$ Å (i.e. $\lambda^{-1} \sin \theta > 0.5$ Å^{-1}, precise crystallographic studies of small molecules have shown that the X-ray model must be much more complex than the neutron model: in both cases, all identifiable nuclear or atomic sites (including hydrogen) are assigned the usual positional parameters, an occupation factor, and at least isotropic thermal parameters. But whereas the bonded electron density (that found in the so-called X–N difference maps) need not be described at all in the neutron model, it must be included in any model that is to predict accurately the X-ray intensities. Conversely, it is possible to determine bonding electron density only from the X-ray data; neutron data are then very useful to give unbiased estimates of the nuclear positions and thermal vibration parameters.

In view of the fact that many protein hydrogen atoms are already visible in high-resolution X-ray Fourier difference maps, e.g. insulin at 1.2 Å resolution (Sakabe et al. 1978), it is to be expected that in the next few years it will be possible to make joint refinements based on X-ray and neutron data, as is already done for smaller molecules (Duckworth et al. 1969).

Further, bonding electron density should be observable in highly ordered regions of small proteins. Direct observation of shifts of bonding electron density consequent on, e.g. oxidation or binding of different substrate molecules, would be the most impressive confirmation of hypotheses advanced to explain the biological interactions of proteins.

In parallel with such very time-consuming studies, it is important to study by neutron diffraction smaller molecules of biological importance, with their water of crystallization. Neutron studies of a vitamin B_{12} monocarboxylic acid (D. C. Hodgkin & B. T. M. Willis, personal communication), of vitamin B_{12} (J. L. Finney, personal communication) both at about 1 Å resolution, and of an α-cyclodextrin hydrate (Saenger 1979) are already well advanced. Certain β-cyclodextrin complexes, which may serve as models of enzyme action or of membrane diffusion transport are very interesting candidates for neutron diffraction, particularly as several structures have already been determined carefully at about -120 °C by X-ray diffraction (Stezowski et al. 1978). For completeness we mention that the most accurate structural information available on biological molecules comes from the many neutron studies of the individual amino acids, mostly at Brookhaven or Trombay. Neutron structures of three dipeptides and of the nucleotide uridine-5'-phosphate are also available (Sequeira 1979).

Position-sensitive detectors for neutrons

High-resolution data collection for proteins larger than lysozyme or cell dimensions larger than 30 Å would require a year or more of beam time with a single detector. In practice, such large amounts of beam time are not available in open competition. For this reason, neutron area detector systems suitable for protein diffraction spectra are being constructed.

Several such systems operate already. Data from CO sperm-whale myoglobin have been extended from 1.8 to 1.5 Å resolution (Schoenborn & Hanson 1979) by using a 20 cm × 20 cm area detector system with spatial resolution about 3 mm (Schoenborn et al. 1978). A segmented linear detector system (Cain et al. 1976) has given good-quality data to 2.2 Å resolution from a 1.6 mm³ crystal – small by neutron standards – of an inhibited trypsin (Kossiakoff & Spencer 1979).

A single linear position-sensitive detector has been used at a medium-flux reactor (Wlodawer 1979) to record reflexions to 2 Å resolution for ribonuclease A. At the I.L.L., Grenoble, a curved area detector to intercept the diffraction pattern over a solid angle of 4 × 64° is under construction. This detector will have 16 vertical wires and 512 horizontal wires to give a resolution of 0.25° horizontally and 0.125° vertically.

Experience with such detectors, and perhaps also with detectors for time-of-flight modified Laue diffractometers at neutron spallation sources, will revolutionize neutron crystallography. Indeed, a careful theoretical study has already shown that with optimization of experimental parameters, high-resolution neutron data on medium-sized proteins might be obtained in a matter of days (Jauch 1979).

The power of X-ray protein crystallography has increased dramatically in the last few years with the availability of synchrotron sources, area detectors and computer programs for least-squares refinement of very large molecules. Thus armed with very good X-ray starting models and high-resolution neutron data, protein crystallographers will be able to learn much about protein structure that could not be learned without neutrons.

We thank E. Duee and A. C. Nunes for their work on triclinic lysozyme and L. H. Jensen for providing the lysozyme X-ray coordinates.

References (Bentley & Mason)

Artymiuk, P. J., Blake, C. C. F., Grace, D. E. P., Oatley, S. J., Phillips, D. C. & Sternberg, M. J. E. 1979 *Nature, Lond.* **280**, 563–568.

Bacon, G. E. 1975 In *Neutron diffraction*, p. 39. Oxford University Press.

Bentley, G. A., Duee, E. D., Mason, S. A. & Nunes, A. C. 1979 *J. Chim. phys.* **76**, 817–821.

Blake, C. C. F., Johnson, L. N., Mair, G. A., North, A. C. T., Phillips, D. C. & Sarma, V. R. 1967a *Proc. R. Soc. Lond.* B **167**, 378–388.

Blake, C. C. F., Mair, G. A., North, A. C. T., Phillips, D. C. & Sarma, V. M. 1967b *Proc. R. Soc. Lond.* B **167**, 365–377.

Cain, J. E., Norvell, J. C. & Schoenborn, B. P. 1976 *Brookhaven Symp. Biol.* **27** (5), 111–143.

Duckworth, J. A. K., Willis, B. T. M. & Pawley, G. S. 1969 *Acta crystallogr.* A **25**, 482–484.

Finney, J. L. 1979 In *Water: a comprehensive treatise*, vol. 6 (ed. F. Franks), pp. 47–122. New York: Plenum Press.

Imoto, T., Johnson, L. N., North, A. C. T., Phillips, D. C. & Rupley, J. A. 1972 In *The enzymes*, vol. 7, pp. 665–868. New York: American Press.

Jauch, W. 1979 Ph.D. thesis, Technical University of Berlin.

Kossiakoff, A. A. & Spencer, S. A. 1979 In *Abstracts, Summer Meeting, Am. Crystallogr. Assoc., Boston University, Boston, Mass.*, p. 28.

Kurachi, K., Sieker, L. C. & Jensen, L. H. 1976 *J. molec. Biol.* **101**, 11–24.

Moult, J., Yonath, A., Traub, W., Smilansky, A., Podjarny, A., Rabinovich, D. & Saya, A. 1976 *J. molec. Biol.* **100**, 179–195.

Norvell, J. C., Nunes, A. C. & Schoenborn, B. P. 1975 *Science, N.Y.* **190**, 568–570.

Phillips, D. C. 1966 *Scient. Am.* **215** (5), 78–90.

Saenger, W. 1979 *Nature, Lond.* **279**, 343–344.

Sakabe, N., Sakabe, K. & Sasaki, K. 1978 In *Proc. Symp. on Proinsulin, Insulin and C-peptide*, Tōkushima, July 1978.

Schoenborn, B. P. 1969 *Nature, Lond.* **224**, 143–146.

Schoenborn, B. P., Alberi, J., Saxena, A. M. & Fischer, J. 1978 *J. appl. Crystallogr.* **11**, 455–466.

Schoenborn, B. P. & Hanson, J. C. 1979 In *A. C. S. Symposium on Water Structure* (ed. J. A. Rupley & I. D. Kuntz). (In the press.)

Sequeira, A. 1979 In *Proc. International Symp. Biomolecular Structure, Conformation and Evolution*, Madras, January 1978 (ed. R. Srinivasan). Oxford: Pergamon.

Stezowski, J. J., Jogun, K. H., Eckle, E. & Bartels, K. 1978 *Nature, Lond.* **274**, 617–619.

Watenpaugh, K. D., Margulis, T. M., Sieker, L. C. & Jensen, L. H. 1978 *J. molec. Biol.* **122**, 175–190.

Wlodawer, A. 1979 *Acta crystallogr.* (Submitted.)

Diffuse neutron scattering from crystal imperfections

By E. W. J. Mitchell† and R. J. Stewart‡
† *The Clarendon Laboratory, Oxford OX1 3PU, U.K.*
‡ *J. J. Thomson Physical Laboratory,
The University, Whiteknights, Reading RG6 2AF, U.K.*

The most extensively studied type of structural diffuse scattering in recent years has been the small angle scattering centred on 000. Examples are given illustrating the types of investigation that can now be made: precipitation in alloys (*in situ* kinetic studies); fracture of alloys (*in situ* mechanical studies); radiation 'damage' in silicon (correlation with spectroscopic measurements); radiation damage in gallium arsenide (anisotropic effects).

In comparison with small angle neutron scattering, much less work has been done on other types of diffuse scattering. Some possible developments in this field are indicated by the computer simulation of relaxation effects around impurity atoms.

Introduction

Crystal imperfections constitute departures from the perfect periodicity of the ideal crystal and therefore give rise to diffuse scattering, i.e. scattering of radiation of wavelengths and at angles other than those given by the Bragg relation. The application of neutrons to the study of imperfections through diffuse scattering has a number of complementary advantages compared with X-rays or electrons.

One of the most important of these is the much greater penetration of neutrons, which allows thick samples to be used and, especially, allows much longer wavelengths to be used. The thickness for 10% absorption at 1 Å† for neutrons is commonly 1–100 cm compared with 5×10^{-4} to 1×10^{-3} cm for X-rays. Indeed, with neutrons it is fairly routine to use wavelengths of 10 Å such that only diffuse scattering occurs and that Bragg and double Bragg scattering are completely eliminated. This is particularly important for small-angle scattering in which, by using smaller wavelengths, the intensities may be seriously affected by double Bragg scattering. In the study of diffuse scattering, advantage may also be taken of one of the general characteristics of the nuclear scattering of neutrons, that the scattering lengths (b) for nuclei of adjacent atomic number can be quite different, whereas for X-rays they are similar. Advantage of this may be taken in studies of materials such as AlMg, AlMgSi or NiFe alloys. The ability to use long wavelength neutrons has the further advantage that because of their low energy (1 meV), phonon emission processes cannot occur, so that the only thermal diffuse scattering possible arises from phonon absorption, which may be effectively eliminated by cooling.

Early measurements of diffuse scattering, either by the total or the differential cross section, were hampered by lack of intensity. The field has, however, developed substantially in recent years through the use of beams from the High Flux Reactor at the Institut Laue–Langevin (I.L.L.) in Grenoble. For small-angle scattering the impact of the I.L.L. has allowed experiments to be done that were previously quite impossible.

† $1 \text{ Å} = 10^{-10} \text{ m} = 10^{-1} \text{ nm}$.

It is convenient to consider three régimes of diffuse scattering, although the division is to some extent arbitrary:

(a) *small angle scattering*, scattering in the vicinity of the reciprocal lattice point (000) produced by relatively large fluctuations in scattering length density;

(b) *diffuse scattering*, between Bragg peaks and between the lowest angle Bragg peak and the straight through beam, produced by strain around defects especially from the relatively large displacements in the first few shells of atoms around the defect;

(c) *Huang scattering*, close to Bragg peaks and difficult to separate quantitatively, produced especially by the long-range part of the strain field around defects.

The requirement to work near Bragg peaks in Huang scattering means that there is no general advantage in using neutrons in that case and we shall not discuss Huang scattering in the present paper.

In this paper we give examples of small-angle neutron scattering (s.a.n.s.) obtained by ourselves and collaborators. The examples illustrate two aspects of s.a.n.s. work: (i) obtaining information about the size and shape of inhomogeneities in crystals, and (ii) using s.a.n.s. to characterize inhomogeneities and to follow their changes with time under various conditions. We then discuss the problem of determining the strain field around point defects, or small complexes, by a combination of diffuse scattering measurements and the computer simulation of models of the displacements of the surrounding atoms.

2. Experimental

(a) Principles

Scattering conditions are described by completing the scattering triangle in reciprocal space. All information about the disorder is contained in the disorder structure factor, $S(Q)$, between reciprocal lattice points, and this has to be determined experimentally. In s.a.n.s. one is determining $S(Q)$ in the vicinity of (000), although similar information is contained around other reciprocal lattice points. However, by working with long wavelengths we eliminate double Bragg scattering processes, which also produce contributions to the intensity at small angles. Our problem, therefore, is to map out the intensity by using scattering triangles within the first set of reciprocal lattice points.

The experiments reported here were performed at the I.L.L. with the instrument D11 which covers both the small-angle range, corresponding to $5 \times 10^{-4} < Q < 5 \times 10^{-1}$ Å^{-1}, and also, with the new counter bank, the range $0.3 < Q < 3$ Å^{-1}. The wavelengths available are from 4.5 to 20 Å. This means that $S(Q)$ may be continuously sampled within the volume of the first Brillouin zone without moving the specimen.

(b) Instrumental

There are currently about ten spectrometers throughout the world suitable for diffuse neutron scattering studies, including two recently commissioned in the U.K. specifically for the small Q region: the Pluto S.A.S. at Harwell and the Herald S.A.S. at Aldermaston.

For work requiring the highest sensitivity in the small Q region, however, D11 at the I.L.L. is by far the best instrument available for all disciplines, so that the time available for a given experiment is severely limited.

D11 has been described fully by Ibel (1976). Neutrons from a cold source, which enhances the long wavelength part of the spectrum, are transmitted through 60 m of curved guide tube into an experimental hall. Being remote from the reactor, this hall has a low background. The wavelength spread ($\Delta\lambda$) in the beam is then limited by a helical velocity selector of resolution ($\Delta\lambda/\lambda$) of 9%. The transmitted wavelength can be varied within the range $4.5 \leqslant \lambda \leqslant 20$ Å by varying the rotational velocity of the selector. The beam then passes through a series of guides and/or apertures which allows the angular divergence to be varied. Thus a collimated beam of long wavelength neutrons is produced. Neutrons scattered by the sample are detected by two arrays of detectors: the small-angle multi-detector and the high-angle detector bank.

The small-angle detector is a $^{10}BF_3$ (1 atm) proportional counter giving a two-dimensional sensitivity by incorporating a multiwire array. It provides 3808 active elements each of 1 cm² spread over a square of 64 cm × 64 cm. The detector can be placed at several distances between 1 and 40 m from the sample, giving an angular range of 18–0.15°. At the centre of the array there is a cadmium beam stop to absorb the unscattered beam.

The high-angle bank, which was installed in 1977, comprises 32 ^{3}He proportional counters, each having a sensitive area of 30 cm × 2.5 cm, arranged at angles between 20 and 140° from the beam line and with constant Q increments. Unlike the small-angle counter, the high-angle bank covers only one portion of the azimuth. Each counter is provided with a collimator to avoid counting stray neutrons.

The entire neutron path from the velocity selector to the detectors is enclosed by one vacuum chamber 80 m long, which is evacuated to 10^{-2} Torr (*ca.* 1.33 Pa) to reduce air scattering.

Finally, we refer to the diffuse scattering spectrometer, which a group associated with one of us (Stewart 1978) has designed for the Spallation Neutron Source. By an ingenious arrangement of movable, annular, position-sensitive detectors the azimuth is fully covered for $10^{-3} \leqslant Q \leqslant 3$ Å^{-1}. Use of the time-of-flight method means that all required wavelengths can be sampled concurrently and that double Bragg scattering and inelastic effects may be discriminated against.

Instrumentally, therefore, the present and future is healthy and we give in the following sections an account of some of the results that we have obtained, together with an indication of future possibilities.

3. SMALL-ANGLE NEUTRON SCATTERING

(a) *Theory*

The theory of small-angle scattering has been reviewed by many authors (e.g. Guinier & Fournet 1955) from various points of view. We give here the essential results for our discussion.

S.a.n.s. arises from fluctuations in the scattering length density $c(\mathbf{r})$ associated with a number density of nuclei $\rho(\mathbf{r})$. The scattering cross section is given by

$$d\sigma/d\Omega = |\mathscr{F}\{c(\mathbf{r})\}|^2, \qquad (1)$$

where $\mathscr{F}$ denotes the Fourier transform. We are especially interested in applying this to an inhomogeneity ($c(\mathbf{r}) = c_i(\mathbf{r})$) in an otherwise homogeneous matrix ($c(\mathbf{r}) = c_m$). For a particle of shape $g(\mathbf{r} - \mathbf{r}_0)$, such that $g_i = 1$ for $\mathbf{r} - \mathbf{r}_0 < r_i$ and 0 for $\mathbf{r} - \mathbf{r}_0 > r_i$, where r_i denotes the boundary measured from the centre of the inhomogeneity $\mathbf{r}_0$, we can write

$$c(\mathbf{r}) = c_m\{1 - g(\mathbf{r} - \mathbf{r}_0)\} + c_i(\mathbf{r}) \cdot g(\mathbf{r} - \mathbf{r}_0).$$

It follows that
$$d\sigma/d\Omega = (\bar{c}_i - c_m)^2 |\mathscr{F}\{g(\mathbf{r}-\mathbf{r}_0)\}|^2, \qquad (2)$$

where
$$\bar{c}_i = \frac{1}{V_i} \int_{r_i} c_i(\mathbf{r}) \cdot d^3\mathbf{r}, \qquad (3)$$

in which V_i is the volume of the inhomogeneity. It is clear from (2) that unless there is a compositional change (different combination of the scattering lengths – b's) or a local macroscopic density change, the s.a.n.s. will be zero.

We may write $\bar{c}_i$ as
$$\bar{c}_i = \frac{1}{V_i} \int_{r_i} (\rho_{ia}(\mathbf{r}) \cdot b_a + \rho_{ib}(\mathbf{r}) \cdot b_b)\, d^3\mathbf{r}$$

for a diatomic system, or for one species
$$\bar{c}_i = \frac{b}{V_i} \int_{r_i} \rho_i(\mathbf{r}) \cdot d^3\mathbf{r}.$$

The method of moments may then be applied to the $\mathscr{F}\{g(\mathbf{r}-\mathbf{r}_0)\}$ in (2) and various combinations of the observed quantities used to calculate characteristic lengths (first moment), areas (second) or volumes (third). Ideally, the observed ratios between these quantities give information about the shape of the particle by comparison with ratios calculated for particular shapes.

The well known equations derived by Guinier (Guinier & Fournet 1955) and Porod (1951) effectively use this approach. For low concentrations N_i (cm^{-3}) of identical, randomly oriented inhomogeneities, the Guinier approximation equation (2) becomes

$$\frac{d\sigma}{d\Omega} = N_i(\bar{c}_i - c_m)^2\, V_i^2\, e^{-\frac{1}{3}Q^2 R_g^2}, \qquad (4)$$

where R_g^2 is the second moment given by

$$R_g^2 = \int_{r_i} \rho(\mathbf{r}) \cdot \mathbf{r}^2\, d^3\mathbf{r} \Big/ \int_{r_i} \rho(\mathbf{r}) \cdot d^3\mathbf{r}, \qquad (5)$$

and in the special case of spheres (R_s), $R_g = (\tfrac{3}{5})^{\frac{1}{2}} R_s$. In Porod's approximation, the cross section is evaluated to the $(QR)^2$ term in the sine expansion, and within that limit the cross section for large Q becomes

$$\frac{d\sigma}{d\Omega} = N_i(\bar{c}_i - c_m)^2 \frac{2\pi}{Q^4} A_i, \qquad (6)$$

where A_i is the surface area of an inhomogeneity. These equations apply to polycrystalline samples and can be used only indirectly to give information about the shape of an inhomogeneity.

More direct information about the shape of the inhomogeneities may be obtained from single crystal measurements. In this case, non-spherical inhomogeneities may be aligned along a few equivalent crystallographic directions and produce anisotropic scattering. This may be calculated by summing over these equivalent directions so that

$$\frac{d\sigma}{d\Omega} = N_i(\bar{c}_i - c_m)^2 \sum_{\langle hkl \rangle} \left| \int_{r_i} g(\mathbf{r}-\mathbf{r}_0) \cdot e^{i\mathbf{Q}\cdot\mathbf{r}}\, d^3\mathbf{r} \right|^2. \qquad (7)$$

For higher concentrations such that interference between the particles has to be included, (2) becomes

$$\frac{d\sigma}{d\Omega} = N_i(\bar{c}_i - c_m)^2 |\mathscr{F}\{g(\mathbf{r}-\mathbf{r}_0)\}|^2 \left[\frac{1}{N_i} \sum_{ij}^{\infty} e^{i\mathbf{Q}\cdot(\mathbf{r}_{0i}-\mathbf{r}_{0j})} \right], \qquad (8)$$

where r_{0i} and r_{0j} are the centres of the inhomogeneities i and j. For widely spaced inhomogeneities the term in the square bracket (the interference term) is 1; for higher concentrations it oscillates but the first maximum is the only one that need be considered.

For randomly distributed inhomogeneities the interference term becomes

$$\{1 - \phi(Q, d)\} \quad \text{with} \quad \phi(Q, d) = 3(\sin Qd - Qd \cos Qd)/(Qd)^3,$$

and a maximum occurs in curves of $d\sigma/d\Omega$ against Q, at

$$Q_{\max} d = 5.76. \tag{9}$$

for example, $d = 100$ Å, $Q_{\max} = 5.76 \times 10^{-2}$ Å^{-1}. In all cases, the presence of a size distribution of the inhomogeneities may smear out the maximum in the scattering cross section.

(b) *Application to alloys*

(i) *Introduction*

The strength of alloys is generally improved by the development of a second phase dispersed throughout the host matrix, often in the form of a precipitate. The strengthening is influenced by factors such as the size distribution of the particles, their concentration and their coherence with the matrix. These factors contribute to diffuse scattering, especially in the low Q region, and by using neutrons we can study thick samples.

(ii) *Aluminium–zinc alloys*

From previous work (see, for example, Kostorz 1979) it appears that the decomposition phase, produced by ageing at relatively low temperatures after quenching from relatively high ones, consists of more or less spherical Zn-rich precipitates. Gerold (1961) has interpreted the X-ray data as indicating spherical precipitates of a single composition, with the observed scattering peak related to interparticle interference. On the other hand, Guinier (1959) has suggested that the effects arise from an inner Zn-rich zone surrounded concentrically by an outer Zn-depleted region.

There are two processes by which such precipitates may be formed: spinodal decomposition, and nucleation followed by growth. The thermodynamic data about the spinodal line are sparse and there have been attempts to use other data to uniquely assign one or other mode of decomposition. Much of our early work was in an attempt to resolve these problems (Allen et al. 1976).

The object of the further experiments that are described here was to follow the decomposition process 'in situ' on D 11 and to see whether the subsequent dissolution process could be similarly followed. A special high-temperature cell was built for incorporation into the beam line; this has been described by Allen (1978).

In earlier experiments, Messoloras (1974) and Allen et al. (1976) found that for compositions of 4.5, 6.7 and 11.8 at. % Zn there were two regions of precipitate growth. The initial fast region was over in $t \leqslant 100$ min, whereas in the second stage growth occurred indefinitely, although slowly. The transition between the régimes was found to be independent of the Zn concentration but was a function of the homogenization temperature (T_Q) from which the samples were quenched.

Results from an 'in situ' experiment are shown in figure 1. Those measurements by Allen (1978) were all for the 11.8 at. % Zn alloy and are given for growth temperatures, T_g, of 110 and 175 °C. T_Q was 300 °C in all cases. Growth temperatures were studied for $100 < T_g < 200$ °C.

The development of the small-angle scattering can clearly be followed: the whole set of curves was obtained in 76 min for $T_g = 110$ °C and 220 min for 175 °C. Guinier plots gave for the end of this initial growth $R_g \approx 34$ Å for $T_g = 110$ °C and 90 Å for 175 °C. From the position of the maxima, and using (4), we find the 34 Å precipitates were separated on average by *ca.* 260 Å while the larger precipitates were *ca.* 770 Å apart. We were thus able to follow the production of well separated larger particles at the higher growth temperature, compared with the closer (260 Å) smaller (34 Å) particles produced at 110 °C.

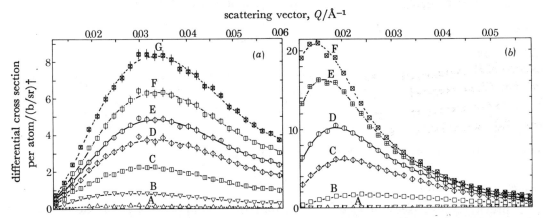

FIGURE 1. Decomposition of an Al–11.8 at. % Zn alloy. (a) Ageing temperature 110 °C; ageing times in minutes: A, 20; B, 27; C, 34; D, 40; E, 46; F, 56; G, 76. (b) Ageing temperature 175 °C; ageing times in minutes: A, 18; B, 39; C, 71; D, 104; E, 163; F, 220.   † 1 barn (b) = 10^{-28} m².

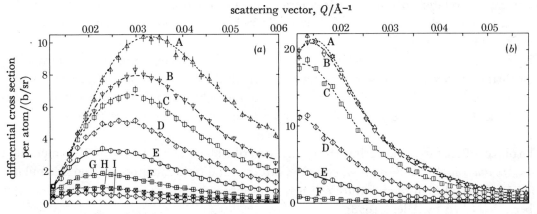

FIGURE 2. Dissolution at 300 °C of an Al–11.8 at. % Zn alloy. (a) Following growth at 110 °C, dissolution times in minutes: A, 0; B, 4.5; C, 7; D, 9; E, 11.5; F, 13.5; G, 16; H, 18; I, 21. (b) Following growth at 175 °C, dissolution times in minutes: A, 0; B, 0.6; C, 1.7; D, 2.8; E, 4.0; F, 5.1.

In figure 2 we show the effect of heating to 300 °C on the precipitates produced under the two growth conditions discussed above. This is the homogenization temperature, and the dissolution of the precipitates – the decrease of the small-angle neutron scattering – is clearly shown. For the sample aged at 175 °C the dissolution was followed and was complete in 7 min. We conclude from these experiments that faster dissolution is associated with higher T_g and therefore that large, widely spaced particles dissolve more rapidly than smaller, more closely packed ones.

We are not able to present a full kinetic analysis here, but the experiments show that such processes may now be readily followed by neutron scattering. We have discussed elsewhere (Allen et al. 1978) the reasons why the above conclusion seems to eliminate intra-particle diffusion as the limiting process in the dissolution.

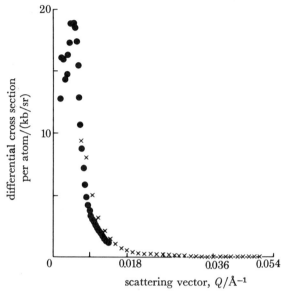

FIGURE 3. Small-angle scattering curve for a Nimonic 105 alloy. Combined 20 m and 5 m sample-detector measurements on D11.

(iii) *Nickel-based 'super-alloys'*

These alloys are composed primarily of Ni, Co, Cr, Mo, Al and Ti. Because of their resistance to creep at high temperatures, they are extensively used in high temperature–stress applications. The resistance to creep is believed to be associated with the size, distribution and concentration of a precipitated phase (the γ' phase). This phase is an ordered alloy having an f.c.c. structure of the form $(NiCo)_3(AlTi)$. The particular alloy examined in the experiments reported here, Nimonic 105, has the atomic composition 52Ni, 20Co, 15Cr, 4Mo, 8Al, 1Ti.

In spite of the good high-temperature creep-resistant properties of these alloys, continuous use at high temperatures and stresses leads to a degradation. It has been suggested that Ostwald ripening of the precipitates is the cause of the degradation, i.e. the dissolution of small precipitates in the vicinity of large ones, which then become larger. The atomic numbers of the main constituents are Al, 12; Ti, 22; Cr, 23; Co, 26; Ni, 27 such that there is relatively low discrimination of the γ' phase in X-ray scattering. The neutron scattering lengths are: Al, 0.35; Ti, -0.34; Cr, 0.35; Co, 0.25; Ni, 1.03. Rearrangement of the Ni, Co, Al, Cr and Ti concentrations therefore gives rise to substantial fluctuations in neutron scattering length density.

Two sets of experiments were carried out, one in which the scattering was measured after an ageing (strengthening) heat treatment and the second in which the scattering was followed dynamically in a high-temperature creep experiment. The latter was carried out in the neutron beam of D 11 in a specially designed cell described by Miller et al. (1978).

For the first experiment the alloy was homogenized by heating at 1150 °C for 16 h and then aged for 16 h at 850 °C. The s.a.n.s. is shown in figure 3. Preliminary analysis showed that in the expected Guinier region the ln $(d\sigma/d\Omega)$–Q^2 plots were curved, and Miller et al. (1978) have used the distribution discussed by Shull & Roess (1947) to derive the distribution of particle sizes. The mean size found was $\bar{R}_g = 300$ Å and hence $\bar{R}_s = 390$ Å.

The sample was then heated for $7\frac{1}{2}$ h at 750 °C under a uniaxial stress of 300 MPa and it was found that Q_{max} has decreased and $\bar{R}_g$ increased by 15%. The increase in both D and $\bar{R}_g$ indicates that some particles have disappeared and others have grown; this is consistent with the Ostwald ripening process.

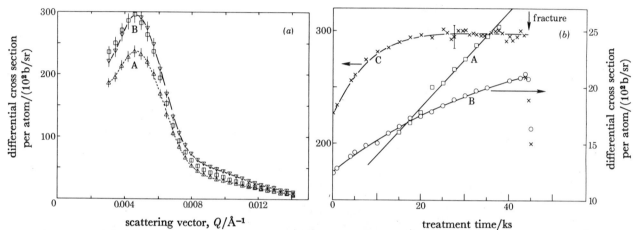

FIGURE 4. (a) Nimonic 105 alloy (A) as heat treated and (B) after 13 h *in situ* treatment at 500 MPa and 800 °C. (b) Comparison of creep (A) and scattering cross section at a Q of 0.014 Å^{-1} (B) and at a Q of 0.005 Å^{-1} (C).

In the dynamic experiment, we measured the creep and neutron scattering simultaneously and wished to monitor both to the point of rupture. In order to achieve this in the available time we used a higher stress and temperature (500 MPa, 800 °C) but after 13 h these were increased to 550 MPa and 850 °C, after which rupture occurred after 30 min. Adequate statistical accuracy was obtained on measurements lasting 250 s. The creep was monitored by a high-precision electrical transducer fixed to the sample and showed a mean fractional creep rate initially of 0.05% h^{-1}.

The scattering pattern at the beginning of the experiment and just before rupture are shown in figure 4a; data taken at $Q = 0.004$ and 0.009 Å^{-1} respectively are compared with the creep rate in figure 4b. While in the early stages of creep both the high and low Q correlate with the creep rate, as the rupture point is approached figure 4b shows that the creep correlates with the higher Q data. The curves in figure 4a show that Q_{max} did not change, unlike the lower temperature–stress experiments. We suggest that in the accelerated stage, as rupture is approached, a further source of scattering in the high Q region develops. One possibility is the formation of microvoids which become the dominant mode of failure in the higher temperature–stress conditions.

Clearly, the process is complicated and further experiments are needed. What has been shown, however, is that these processes may be followed on thick samples by '*in situ*' s.a.n.s.

(c) *Application to irradiated semiconductors*

(i) *Introduction*

When crystals are irradiated with high energy particles, atoms of the crystal are displaced from their normal position. Various point defects (vacancies, interstitials) or small defect complexes (divacancies, di-interstitials, vacancy–impurity complex, interstitial–impurity complex, etc.) may be formed. At one extreme (e.g. bombardment with 0.5 MeV electrons), the recoil energies are small (less than 100 eV, say) and the process is close to threshold. In the other extreme (e.g. fission energy neutrons in a reactor), the recoil energies are *ca.* 0.5 MeV and the energetic knock-on atoms produce *ca.* 1000 further displacements in coming to rest.

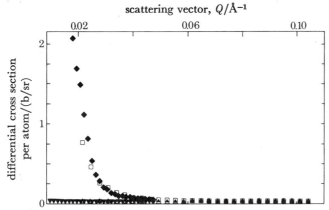

FIGURE 5. The small-angle scattering from irradiated ($\square$, $\blacklozenge$) and unirradiated ($\triangle$, $\blacktriangledown$) GaAs. $\triangle$, $\square$, 2 m data; $\blacktriangledown$, $\blacklozenge$, 5 m data. Irradiation was with 9×10^{19} neutrons/cm².

Because of the penetration of the neutron, this damage by energetic knock-on atoms is produced throughout the crystal. However, in a number of semiconductors there is indirect evidence from the electrical and optical properties of the irradiated crystals that at high neutron doses the damage is not uniformly distributed (e.g. silicon (Gossick 1959), gallium arsenide (Coates & Mitchell 1975)). In the present experiments we have used s.a.n.s. to obtain structural information above the defect distribution. If the defects occur in clusters, for example, there should be a density fluctuation that would give rise to s.a.n.s.

(ii) *Gallium arsenide*

Coates & Mitchell (1975) have described the changes in electrical and optical properties that follow the introduction of defects into GaAs by fast neutron irradiation. From the analysis of these results they concluded that the defects were not uniformly distributed. Subsequently Gupta *et al.* (1978) carried out s.a.n.s. measurements.

The s.a.n.s. curves for irradiated and unirradiated crystals are shown in figure 5. There is pronounced small-angle scattering with $R_g \approx 150$ Å. However, the scattering was not isotropic and the scattering pattern for the incident neutron beam along [111] is shown in figure 6.

This pattern has been compared with computations based on (7). The anisotropy in the scattering pattern is that of the Fourier transform of the shape function of the inhomogeneity. Gupta *et al.* (1978) concluded that the fluctuations were ellipsoidal. Further work is in progress concerning both the orientation of the ellipsoids and the method of their production by the radiation damage process.

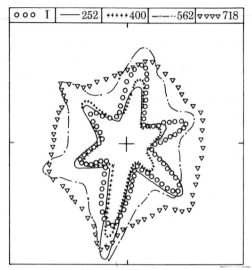

FIGURE 6. The effects of annealing on the anisotropic scattering cross section of irradiated GaAs. (Temperature in degrees Celsius.)

(iii) *Silicon*

Beddoe *et al.* (1979) have measured the s.a.n.s. from fast neutron irradiated ultra-purity polycrystalline silicon obtained from the Dow Corning Corporation. The form of the scattering is shown in figure 7a and the corresponding Guinier analysis in figure 7b. There seem to be two characteristic sizes associated with the scattering, *ca.* 150 Å dominating the lower Q scattering and *ca.* 50 Å the higher Q.

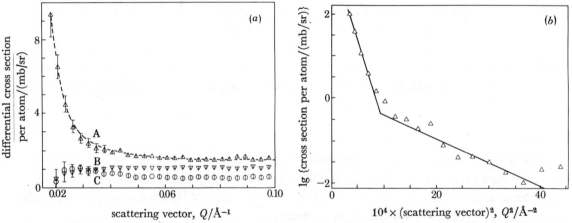

FIGURE 7. (a) The irradiation-induced cross section in a polycrystalline silicon sample irradiated with 10^{20} fast neutrons (A); after annealing at 110 °C (B) and after annealing at 700 °C (C). (b) Guinier plot of the irradiation induced cross section.

It was possible to study the annealing of this scattering and two of the curves are shown in figure 7a. It is clear that substantial annealing occurred at 100 °C throughout the whole Q region, but especially at low Q.

From the e.s.r. measurements of Watkins & Corbett (1965) and the infrared measurements of the 1.7 μm band by Cheng *et al.* (1966), it is known that divacancies are the main defects in room-temperature electron or neutron irradiated silicon. The divacancies are found to anneal between 100 and 200 °C, the temperature being found by Newman & Totterdell (1975) to be lower for the higher dose irradiations. We propose, therefore, that the s.a.n.s. arises from regions of crystal ($R \approx 150$ Å) containing a high concentration of divacancies. The s.a.n.s. remaining after the 100 °C annealing was further decreased by heating at 700 °C, and this may well be associated with the spectroscopic effects, possibly due to higher vacancy complexes, which were found by Koval et al. (1973) and Newman & Totterdell (1975) to anneal at 600 °C.

An interesting aspect of this experiment is the correlation of the structural data (s.a.n.s.) with the spectroscopic data through the annealing characteristic.

4. Diffuse neutron scattering

(a) Theory

There is a considerable body of theoretical work on the calculation of the form of the diffuse scattering produced by various kinds of defect. The calculations are developments to suit particular purposes of the methods used by James (1948), but can now be carried much further because of the computational power that is available. Good examples of detailed calculations are in the work of Martin (1965), Stewart (1969), Schmatz (1973), Clark et al. (1971) and Gupta (1976).[†]

The theory of diffuse scattering has been reviewed by Schmatz (1973) and discussed extensively by Krivoglaz (1969) in his monograph. These include accounts of the formalisms in which the displacements are described by the virtual forces of the lattice statics developed by Kanzaki (1957). We refer to the convenience of this approach later. It has the disadvantage, apart from the conceptual difficulty of the understanding of fictitious forces as opposed to atomic displacements, of only being applicable for small displacements. There is evidence, however, from the calculations of Larkins & Stoneham (1971) and the experiments of Clark et al. (1971), that the displacements of the first two or so shells of atoms around a defect are not necessarily small. In this paper, therefore, we give the essential results of the method that we have used, which provides a description of diffuse scattering in terms of displacements.

We consider an incident beam of amplitude ψ_i and wavevector $\mathbf{k}_i$ incident on a crystal in a direction α with respect to the crystal axes. The beam scattered by N_d (cm^{-3}) defects, each of scattering length b_d, at an angle ϕ has amplitude $\psi_{\mathrm{fd}}(\alpha, \phi)$ and wavevector $\mathbf{k}_f$. The atoms of the host (N cm^{-3}) have scattering length b; for a vacancy $b_d = -b$, for a substitutional impurity atom (b_I), $b_d = b_I - b$, and similarly for small complexes. We describe the normal[‡] lattice positions as $\mathbf{R}_l$ and suppose that the defects are at various $\mathbf{R}_L$. Around each defect (we refer to these as ca. $\mathbf{R}_L$) we consider atoms displaced by $\mathbf{u}_l$ from their positions $\mathbf{R}_{L+l}$ in the reference lattice. The scattered amplitude may then be represented by

$$\psi_{\mathrm{fd}}(\alpha, \phi) = A(\mathbf{R}_l) + B(\mathbf{R}_L) + C(\mathbf{R}_{L+l} + \mathbf{u}_l) - D(\mathbf{R}_{L+l}), \tag{10}$$

comprising contributions from all reference lattice sites (A), all defects (B), the displaced atoms around each defect (C) and the sites from which these atoms have been displaced (D). From

---

[†] Other less detailed calculations are in the work of Antal et al. (1958), Sabine et al. (1962), Mitchell (1966) and Belson (1968).

[‡] It makes little difference for our purposes whether we consider the average lattice of the crystal with defects as the reference or the unperturbed lattice; strictly it should be the former.

(10) the calculations are straightforward in principle but complicated in practice. The full expression is

$$\psi_{\text{fd}}(\boldsymbol{\alpha}, \phi) = \sum_{l=0}^{N} b\, e^{-i\boldsymbol{Q}\cdot\boldsymbol{R}_l} + \sum_{L=0}^{N_{\text{d}}} \left\{ b_{\text{d}}\, e^{-i\boldsymbol{Q}\cdot\boldsymbol{R}_L} + \sum_{\substack{ca.\ L \\ l\neq L}}^{N} b\, e^{-i\boldsymbol{Q}\cdot(\boldsymbol{R}_{L+l}+\boldsymbol{u}_l)} - \sum_{\substack{ca.\ L \\ l\neq L}}^{N} b\, e^{-i\boldsymbol{Q}\cdot\boldsymbol{R}_{L+l}} \right\}, \quad (11)$$

which may be simplified to

$$\psi_{\text{fd}}(\boldsymbol{\alpha}, \phi) = \sum_{l=0}^{N} b\, e^{-i\boldsymbol{Q}\cdot\boldsymbol{R}_l} + \sum_{L=0}^{N_{\text{d}}} e^{-i\boldsymbol{Q}\cdot\boldsymbol{R}_L} \left\{ b_{\text{d}} + \sum_{\substack{ca.\ L \\ l\neq L}}^{N} b\, e^{-i\boldsymbol{Q}\cdot(\boldsymbol{R}_l+\boldsymbol{u}_l)} - \sum_{\substack{ca.\ L \\ l\neq L}}^{N} b\, e^{-i\boldsymbol{Q}\cdot\boldsymbol{R}_l} \right\}. \quad (12)$$

In (12), the term in braces represents the contribution to the scattering of one defect and its associated strain field. The summation over the random sites $\boldsymbol{R}_L$ gives a multiplying factor of $N_{\text{d}}^{\frac{1}{2}}$ and we have

$$\psi_{\text{fd}}(\boldsymbol{\alpha}, \phi) = \sum_{l=0}^{N} b\, e^{-i\boldsymbol{Q}\cdot\boldsymbol{R}_l} + N_{\text{d}}^{\frac{1}{2}} \left\{ b_{\text{d}} + \sum_{\substack{ca.\ L \\ l\neq L}}^{N} b\, e^{-i\boldsymbol{Q}\cdot(\boldsymbol{R}_l+\boldsymbol{u}_l)} - \sum_{\substack{ca.\ L \\ l\neq L}}^{N} b\, e^{-i\boldsymbol{Q}\cdot\boldsymbol{R}_l} \right\}, \quad (13)$$

or, in the form similar to (10),

$$\psi_{\text{fd}}(\boldsymbol{\alpha}, \phi) = \mathscr{A} + N_{\text{d}}^{\frac{1}{2}}\{\mathscr{B} + \mathscr{C} - \mathscr{D}\}. \quad (14)$$

The differential scattering for a single crystal is $(d\sigma/d\Omega)(\boldsymbol{\alpha}, \phi) = N^{-1}|\psi_f \psi_f^*|$; for a polycrystalline sample this is averaged over all $\boldsymbol{\alpha}$; the total cross section $\sigma(|k_i|)$ for such a sample is the latter integrated over all ϕ and Ω; and for a single crystal $(d\sigma/d\Omega)(\boldsymbol{\alpha}, \phi)$ is integrated over all ϕ and Ω to give $\sigma(\boldsymbol{k}_i)$. These quantities are interrelated; in particular it is fairly easily shown that the information for a polycrystalline sample contained by $d\sigma/d\Omega$ as a function of ϕ is equivalent to σ as a function of $|k_i|$, provided $|k_i| \leq \frac{1}{2}\tau_{\min}$, where $\tau_{\min}$ is the minimum reciprocal lattice vector.

To calculate the differential scattering, we consider a matrix whose components are the order of N_{d} of the relevant terms. We have

	$\mathscr{A}$	$\mathscr{B}$	$\mathscr{C}$	$\mathscr{D}$
$\mathscr{A}^*$	0	$N_{\text{d}}^{\frac{1}{2}}$	$N_{\text{d}}^{\frac{1}{2}}$	$-N_{\text{d}}^{\frac{1}{2}}$
$\mathscr{B}^*$	$N_{\text{d}}^{\frac{1}{2}}$	N_{d}	N_{d}	$-N_{\text{d}}$
$\mathscr{C}^*$	$N_{\text{d}}^{\frac{1}{2}}$	N_{d}	N_{d}	$-N_{\text{d}}$
$\mathscr{D}^*$	$-N_{\text{d}}^{\frac{1}{2}}$	$-N_{\text{d}}$	$-N_{\text{d}}$	N_{d}

For typical defect concentrations we need only consider the terms in N_{d}, that is, those in the box. The main computation problem, therefore, is dealing with the displacements around each defect.

Away from the defect (say from the third or fourth shells, but this has to be determined empirically) the displacements will be of the form deduced by Eshelby (1956) from continuum theory:

$$u_l = a/|\boldsymbol{R}_l - \boldsymbol{R}_L|^2. \quad (15)$$

The strength and sign of the dilatation has to be assigned. Larkins & Stoneham (1971) and Clark *et al.* (1971) have considered first shell displacements as large as 20%. We have also to allow for displacements of opposite sign in the first two shells (see §4b). Some simplifications in computing procedure have been made by Gupta (1976).

The alternative analytical approach has been carried furthest by Schmatz (1973) and this introduces considerable simplifications when *all* the displacements are small. When *all* the displacements are *not* small, Schmatz (1973) retains the formalism appropriate to small displacements but introduces a correction term; this may be done exactly but in so doing the same complications re-enter.

In the limit of small displacements,

$$e^{iQ\cdot(R_l+u_l)} \quad \text{becomes} \quad iQu_l\,e^{iQ\cdot R_l},$$

and we may write
$$|\psi_f\psi_f^*| = N_d\,|b_d + b\{iQ\mathscr{F}\{u_l\} + H(u_l)\}|^2, \tag{16}$$

where the Fourier series
$$\mathscr{F}\{u_l\} = \sum_{l=-\infty}^{\infty} u_l\,e^{iQ\cdot R_l}.$$

The correction term $H(u_l)$ if the displacements are not small is

$$H(u_l) = \sum_{l=-\infty}^{\infty} e^{iQ\cdot R_l}\,(e^{iQ\cdot u_l} - 1 - iQu_l). \tag{17}$$

Equation (16), which follows directly from (13), is the definitive equation (47) referred to by Schmatz (1973). For very small Q,
$$H(u_l) \to 0$$

and
$$|\psi_f\psi_f^*| \to N_d b_d^2 \quad \text{or} \quad d\sigma/d\Omega = c_d b_d^2.$$

If the u_l are small, the scattered intensity is related to the Fourier transform of the displacements, which in turn can be represented by the dynamical matrix of the phonon displacements. When they are not small, the use of (17) involves a physically fictitious correction and we prefer to compute directly from (13). We have used various assumed values of real displacements usually for the first two shells (say 20 atoms) with a continuum equation for subsequent shells. Thus the computation is readily managed for three unknowns: first and second shell displacements and a strength parameter for the continuum theory.

(b) Computational studies

The effects of relaxation around a defect on the scattering cross section can be calculated by using the expressions derived in the previous section, provided some assumptions are made to determine the positions of the atoms in the radial strain field produced by the defect. In an elastic continuum the amount of relaxation surrounding a point source of dilation is given by (15). For the first neighbour shell various assumed relaxations can be used to determine the value of the strength parameter a. Radial relaxations for subsequent shells are calculated from (15).

An extensive series of such calculations to determine the total cross section were made by Clark *et al.* (1971) for various defects in germanium. The discrimination that can be achieved in the determination of the details of the relaxation field associated with a defect are illustrated by the calculations of the differential cross section arising from various relaxation models associated with a divacancy in a single crystal of silicon (M. R. Baig & R. J. Stewart, personal communication). In these calculations, the relaxations out to the fifth nearest neighbour shell were included, involving a total of 54 atoms. We illustrate the type of effects that it may be possible to observe in the computational results shown in figure 8. Here the first shell is relaxed inwards by 20% and the third, fourth and fifth shells are given inward relaxations of amounts given by (15). The second shell, however, is varied from inward (also given by (15)), to no relaxation and to outward relaxation. Such opposite relaxations of second shells of atoms surrounding defects have been shown to occur in lattice calculations involving interatomic force simulations.

The results in figure 8 show that the second shell position considerably influences $(d\sigma/d\Omega)$ for $Q \approx 0.1$, 0.28 and 0.35 Å^{-1}. Such differences should be observable. It is clearly desirable that diffuse scattering measurements are made simultaneously throughout the whole Q region as is now possible with D 11 and will be possible with the S.N.S. small-angle scattering instrument.

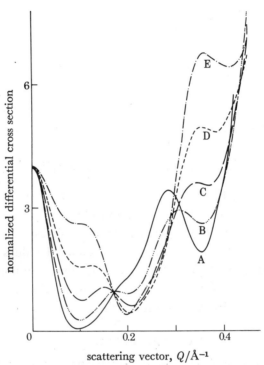

FIGURE 8. Theoretical differential cross section for divacancies in silicon, with the first nearest neighbour atoms relaxed towards the divacancy by 20%. The next four nearest neighbour shells are relaxed according to (15) with the exception of the second nearest neighbours, for which the relaxation is modified. The values of a in (15) for this are: a for curve A, $\frac{1}{2}a$ for B, 0 for C, $-\frac{1}{2}a$ for D and $-a$ for E.

(c) Experiments

Very little experimental work has so far been carried out. Clark *et al.* (1971) compared the λ dependence of the total cross section with various relaxations around defects in germanium. The most interesting experiment, however, is that of Seitz *et al.* (1975), who measured the diffuse scattering of single crystals of Pb containing 2% and 4% Bi. The experiments show the effects of the relaxation of Pb atoms surrounding the impurity. Calculations for this system have been made by Schumaker *et al.* (1973). The experimental results are in good agreement with a strain of trigonal symmetry. The detailed intensities in the 100, 110, 111 representative triangle are well described by the relaxation model provided that both attractive and repulsive (fictitious – see §4a) forces are assumed.

The possibilities of determining these strain effects, about which there is very little information, from diffuse neutron scattering looks extremely promising.

We should like to thank colleagues who have collaborated with us over a number of years in this work at the University of Reading, A.W.R.E. (Aldermaston) and the Institut Laue-Langevin (Grenoble).

BIBLIOGRAPHY (Mitchell & Stewart)

Allen, D. R. 1978 Ph.D. thesis, University of Reading.
Allen, D. R., Epperson, J. E., Gerold, V., Kostorz, G., Messoloras, S. A. & Stewart, R. J. 1976 In *Proc. Conf. on Neutron Scattering* (ORNL Report no. CONF-760601-P1), pp. 102–108.
Allen, D. R., Messoloras, S. A., Stewart, R. J. & Kostorz, G. 1978 *J. appl. Crystallogr.* **11**, 578–580.
Antal, J. J. & Goland, A. N. 1958 *Phys. Rev.* **112**, 103–111.
Beddoe, R. E. 1978 Ph.D. thesis, University of Reading.
Beddoe, R. E., Messoloras, S. A., Mitchell, E. W. J. & Stewart, R. J. 1979 *Inst. Phys. Conf. Ser.* no. 46, pp. 258–266.
Belson, J. 1968 Ph.D. thesis, University of Reading.
Cheng, L. J., Corelli, J. C., Corbett, J. W. & Watkins, G. D. 1966 *Phys. Rev.* **152**, 761–774.
Clark, C. D. & Meardon, B. H. 1972 *Nature, phys. Sci.* **235**, 18–20.
Clark, C. D., Messoloras, S. A., Mitchell, E. W. J. & Stewart, R. J. 1975 *J. appl. Crystallogr.* **8**, 127.
Clark, C. D., Mitchell, E. W. J. & Stewart, R. J. 1971 *Cryst. Lattice Defects* **2**, 105–120.
Coates, R. & Mitchell, E. W. J. 1975 *Adv. Phys.* **24**, 593–644.
Eshelby, J. D. 1956 *Solid State Phys.* **3**, 79–144.
Gerold, V. 1961 *Physica Status Solidi* **1**, 37–49.
Gossick, B. R. 1959 *J. appl. Phys.* **30**, 1214–1218.
Guinier, A. 1959 *Solid State Phys.* **9**, 293–398.
Guinier, A. & Fournet, G. 1955 *Small angle scattering of X-rays*. New York: Wiley.
Gupta, S. 1976 Ph.D. thesis, University of Reading.
Gupta, S., Mitchell, E. W. J., Stewart, R. J. & Kostorz, G. 1978 *Phil. Mag.* A **37**, 227–243.
Hosemann, R. & Bagchi, S. N. 1962 *Direct analysis of diffraction by matter*. Amsterdam: North-Holland.
Ibel, K. 1976 *J. appl. Crystallogr.* **9**, 296–309.
James, R. W. 1948 *Optical principles of the diffraction of X-rays*. London: G. Bell.
Kanzaki, H. 1957 *J. Phys. Chem. Solids* **2**, 24–37.
Kostorz, G. 1979 In *Treatise on materials science and technology* (ed. G. Kostorz), vol. 15, pp. 227–289. New York: Academic Press.
Koval, Y. P., Mordkovich, V. N., Temper, E. M. & Kharchenko, V. A. 1973 *Soviet Phys. Semicond.* **6**, 1152–1155.
Krivoglaz, M. A. 1969 *Theory of X-ray and thermal neutron scattering by real crystals*. New York: Plenum Press.
Larkins, F. P. & Stoneham, A. M. 1971 *J. Phys.* C **4**, 143–163.
Laslaz, G., Kostorz, G., Roth, M., Guyot, P. & Stewart, R. J. 1977 *Physica Status Solidi* a **41**, 577–583.
Martin, D. G. 1965 Ph.D. thesis, University of Reading.
Meardon, B. H. 1970 Ph.D. thesis, University of Reading.
Messoloras, S. A. 1974 Ph.D. thesis, University of Reading.
Miller, R. J. R., Messoloras, S. A., Stewart, R. J. & Kostorz, G. 1978 *J. appl. Crystallogr.* **11**, 583–588.
Mitchell, E. W. J. 1966 *A.E.R.E. Report* no. R5269 230–242. United Kingdom Atomic Energy Authority.
Newman, R. C. & Totterdell, D. H. T. 1975 *J. Phys.* C **8**, 3944–3954.
Porod, G. 1951 *Kolloidzeitschrift* **124**, 83–95.
Powell, M. J. D. 1965 *Comput. J.* **7**, 303–307.
Sabine, T. M., Pryor, A. W. & Hichman, B. S. 1962 *Phil. Mag.* **8**, 43–57.
Schumacher, H., Schmatz, W. & Seitz, F. 1973 *Physica Status Solidi* a **20**, 109–117.
Schmatz, W. 1973 In *Treatise on materials science and technology* (ed. H. Herman), vol. 2, pp. 105–229. New York: Academic Press.
Seitz, E., Schmatz, W., Bauer, G. & Just, W. 1975 *J. appl. Phys.* **8**, 183.
Shull, C. G. & Roess, L. C. 1947 *J. appl. Phys.* **18**, 295–307.
Stewart, R. J. 1969 Ph.D. thesis, University of Reading.
Stewart, R. J. 1978 *The small angle neutron scattering apparatus for the SNS*. Didcot: Rutherford Laboratory.
Walker, C. & Guinier, A. 1953 *Acta metall.* **1**, 568–579.
Watkins, G. D. & Corbett, J. W. 1965 *Phys. Rev.* A **138**, 543–555.

Discussion

R. J. R. MILLER (*Ministry of Defence, Main Building, Whitehall, London, U.K.*). I should like very briefly to describe the small-angle neutron scattering machine at the M.O.D.'s 5 MW reactor HERALD at Aldermaston to give context to my remarks. The machine consists of a 3 l H_2–D_2 cold source at 20 K, giving a cold beam of neutrons which are passed through a conventional velocity selector onto the sample, a 4.5 m evacuated flight tube and a multi-detector. The latter is a 128 × 128 element detector made by the C.E.N.G. The velocity selector has a variable resolution of 5–25% and the collimation before the sample will be done by thin-wall Soller collimators. The machine is very much less powerful (two orders of magnitude) than the D 11 machine at Grenoble, but there is still a wide range of useful experiments that can be performed.

In particular, measurements on technological samples where differences from the norm are sought can be made. Many of these samples are very powerful scatterers of neutrons at small angles, so that very powerful neutron sources are not absolutely necessary. In addition, the mechanical properties of technologically interesting materials, almost by definition, vary slowly with time so that '*in situ*' experiments, where the sample is subjected to typical operating temperatures, stresses, etc., may need to take many weeks. Such long-term experiments can only be carried out on less powerful machines.

E. W. J. MITCHELL. Clearly there are experiments that can usefully be done on medium flux reactors, especially if the long wavelength flux is enhanced by a cold source, as in the example given by Dr Miller. The experiments described in our paper were chosen to illustrate the types of measurements that may be made with the use of high fluxes.

D. J. CEBULA (*Rutherford Laboratory, Didcot, Oxon., U.K.*). Many of the scattering patterns presented display marked peaks in the low Q region. These peaks can presumably be attributed to diffraction effects arising from the particle–particle correlations in the sample. Has any attempt been made to separate the structure factor part of the total signal from the single particle factor and thus obtain quantitative information about the radial distribution of particles?

E. W. J. MITCHELL. In principle, it is possible to obtain information about the radial distribution function of the particles. However, in practice it is difficult to obtain sufficient data on the low Q side of the maximum. Furthermore, unless the particles are closely of the same size, the higher ripples of the interference function are lost.

Magnetic diffuse and small-angle scattering

By W. Schmatz

*Kernforschungszentrum Karlsruhe, Institut für Angewandte Kernphysik,
Postfash 3640, D-7500 Karlsruhe 1, Germany*

Magnetic neutron scattering is one of the most valuable tools in exploring magnetism. For disordered magnetic systems, in addition to Bragg scattering, diffuse and small-angle scattering give information. Experimental techniques of the two last mentioned methods are described. They succeeded in a microscopic explanation of magnetic alloys with paramagnetic, spin-glass, ferromagnetic and antiferromagnetic character. Typical examples will be discussed in detail. Non-frozen local magnetic moments show quasi-elastic scattering, the half-width of which is a characteristic for the relaxation processes. By diffuse inelastic neutron scattering the temperature dependence and concentration dependence of such relaxation processes have been studied for classical paramagnetic, Kondo- and intermediate-valence systems. An additional impact to the field is to be expected from new developments of polarization analysis techniques.

1. Introduction

The number of scattering studies of disordered structures has rapidly increased within the last decade for two reasons: (i) the improvement of experimental techniques, especially at high flux reactors, now allows detection of weak scattering intensities; and (ii) solid state physicists have become more interested in disordered structures. With neutron small-angle and diffuse scattering, information on spatial correlations both on a long-range and a local atomic scale is obtained. This holds for both nuclear and magnetic thermal neutron scattering. Though this lecture deals only with magnetic small-angle and diffuse scattering, the subject is too broad to be discussed in detail. Therefore emphasis will be given to the underlying general aspects, followed by four short specific sections. I have recently reviewed the subject (Schmatz 1978). Only some very recent developments will be referred to here in addition.

2. General aspects

In magnetic neutron scattering, the relevant interaction operator is $\mu_n \cdot B$, where μ_n is the neutron magnetic moment – a product of the absolute value of μ_n and the Pauli spin matrices – and B the operator of magnetic induction of the scattering system. The scattering formalism shows that in a scattering experiment the Fourier transform of the space- and time-correlation of the local magnetic moments is measured as a function of the momentum transfer $\hbar Q$ and the energy transfer $\hbar \omega$ to the scattering system. The scattering is, however, also a function of the initial neutron spin polarization p_0 via $\mu_n B$. For instance, with $p_0 = 0$ no interference terms between nuclear and magnetic scattering appear in the scattering intensity, whereas the intensity can be altered considerably by varying $p_0 \neq 0$ both in direction and absolute magnitude. Even more information can be obtained in magnetic neutron scattering if in addition to the scattering intensity the polarization p_1 of the scattered beam as a function of

Q, ω and p_0 is determined. But, with a few exceptions, this additional source of information is very little explored today even where it would be essential for the problem under investigation. To get p_1 requires considerable effort and it is a real challenge to develop more suitable techniques for the polarization analysis of scattered neutron beams. Possibly the mass production of supermirrors is an essential basis for new devices.

Another general aspect to be considered within our topic is the limit of scattering theory in first Born approximation. This approximation requires that an inhomogeneity will attenuate the traversing beam only by a fraction much smaller than unity. This holds if the total scattering cross section σ_{tot} is much smaller than the geometrical cross section σ_{geom}. Now (assuming a spherical inhomogeneity), it can easily be shown that σ_{tot} increases with the fourth power of the diameter D of the inhomogeneity whereas the geometrical cross section is proportional to D^2. This means for large inhomogeneities ($\sigma_{tot} > \sigma_{geom}$), a much more complicated scattering theory is necessary. Only for very large scattering objects does one return to a simple theory (geometrical optics) again.

For the discussion of the various possibilities of scattering experiments on disordered structures, a subdivision of reciprocal space (Q-space) is reasonable. At first, we have to consider the immediate neighbourhood of $Q = 0$ and $Q = \tau_{hkl}$, where τ_{hkl} is the reciprocal lattice vector for the reflecting plane hkl. Within a radius of 10^{-4} Å^{-1}† around these points large inhomogeneities as grain boundaries, magnetic domain walls and surface inhomogeneities may give scattering intensity in this region, which can be accompanied also by neutron depolarization. The scattering effects are rather strong and thus for $Q = 0$ scattering multiple scattering may occur, which may extend the scattering effects of such large inhomogeneities up to effective scattering vectors of $Q = 10^{-2}$ Å^{-1} in extreme cases. We name this region of variable extension the 'nearby $Q = 0$' region. Small-angle neutron scattering (s.a.n.s.) in its original sense was meant as scattering performed under small scattering angles. We define s.a.n.s. in this context as scattering that can be interpreted in first Born approximation for Q values between 10^{-4} Å^{-1} and 0.3 Å^{-1}. The lower limit is somewhat arbitrarily determined by the best resolution obtainable today. The upper limit is reasonable because for much larger Q values the local atomic arrangements has to be considered in calculating scattering cross sections, whereas in s.a.n.s. theory one prefers to define a density of scattering lengths as an average value for some hundreds of atoms. The nearby $Q = 0$ region as defined above and the s.a.n.s region may overlap, but this makes sense because they are defined for quite different scattering phenomena. The total Q region except the nearby $Q = 0$ (and the nearby $Q = \tau_{hkl}$) region can be regarded as the region of diffuse neutron scattering (d.n.s.). However, having in mind that diffuse scattering, as originated by more or less randomly distributed local atomic fluctuations, is normally very weak, one knows from experimental experience that this region can hardly be explored very near to $Q = 0$ or $Q = \tau_{hkl}$. First, the low intensities require mostly large resolution elements in Q-space ($\Delta Q \approx 0.1$ Å^{-1}) and in the experiment it is absolutely necessary not to include $Q = 0$ or $Q = \tau_{hkl}$ even with the slightest tail of the resolution distribution around the average Q value. Secondly, near $Q = \tau_{hkl}$, the reciprocal lattice point may be extended considerably by azimuthal mosaic spread. And finally, near $Q = \tau_{hkl}$, low frequency phonons may be emitted or absorbed by the incident neutrons with a rather high scattering cross section, which gives a considerable background intensity. Thus diffuse scattering

† 1Å $= 10^{-10}$ m $= 10^{-1}$ nm.

is normally performed at least at a distance of 0.1 Å^{-1} from $Q = 0$ or $Q = \tau_{hkl}$. There are, however, special cases in which it is worthwhile to go very near to the reciprocal lattice points; for instance in studies of critical scattering or if long range lattice distortions are of interest.

The last general aspect to be discussed is the separation of elastic and inelastic scattering. S.a.n.s. and d.n.s. are often performed without any energy analysis of the scattered beam. The 'Q-dependent' scattering pattern is obtained as a 'snapshot' with an aperture opening time of $\hbar/E_0$, where E_0 is the energy of the incident neutron beam. The $|Q|-\omega$ relation for a fixed scattering θ, however, makes it quite uncertain for what average Q value the snapshot has been taken for a given scattering angle. We have to be sure that the scattering is preferentially concentrated around $\omega = 0$ in an interval $\Delta\omega$ much smaller than $E_0/\hbar$. Then Q is, to good approximation, $4\pi\lambda^{-1}\sin\theta$, where λ is the neutron wavelength. Otherwise we have to perform an energy analysis of the scattered beam. In nuclear scattering, the condition above is normally fulfilled rather well for the 'elastic' scattering of disordered structures. (Only in a few cases, e.g. for hydrogen diffusion in metals, does one have to be careful.) Magnetic moments, however, have often short relaxation times for spin diffusion and there are really extreme cases, as we shall see in §6.

3. Zero beam experiments

One of the simplest neutron scattering experiments is given by the transmission of a well collimated beam through a demagnetized slab of magnetically soft iron containing many magnetic domains along the neutron path. The beam is broadened by multiple refraction and this broadening can easily be measured with a detector at a sufficient distance. With slit geometry, even in low flux reactors the intensity for such an experiment is high enough. The most intensive study of a zero beam case has been performed within the last few years by Schärpf (1977). Neutrons enter ferromagnetic nickel with a well defined magnetic domain structure. The solution of the Schrödinger equation in the domain wall is based on four fundamental solutions, which in the general case are linearly combined with four times two coefficients (phases and amplitudes) to make the final neutron wave according to the boundary conditions. The results, e.g. the transmission coefficient, depend sensitively on the Bloch wall parameters. A small-angle scattering apparatus with very high resolution (2″), consisting of two parallel silicon monochromators with the sample between, was used to measure the separation of transmitted and deflected beam. The deflexion angle α (figure 1a) decreases with the tilt angle θ (figure 1b). The qualitative explanation is straightforward: the neutron path in the Bloch wall (thickness 200 Å) decreases with increasing θ.

Depolarization of the zero beam has been used to study characteristic parameters of domain walls and also for the analysis of inhomogeneous magnetization distribution in type II superconductors with pinning forces. The depolarization direction can be reversed rather rapidly with spin flippers. In two recent experiments, transient phenomena in zero beam neutron depolarization have been studied by this promising technique. Rekveldt & van Schaik (1979) observed single domain-walls on motion in a FeSi (3.5% by mass) picture frame crystal. Badurek et al. (1979) also applied this technique for the observation of magnetic after-effects, e.g. in the superparamagnetic alloy Cu–1% Co.

4. Neutron small-angle scattering

A consequent design with regard to luminosity and resolution considerations has led (1968–69) to the construction of the D 11 s.a.n.s. apparatus at the Institut Laue–Langevin (I.L.L.), which still can be regarded as the most powerful instrument of this type in the world. The incident neutron wavelength can be varied between 5 and 20 Å with a wavelength resolution, $\Delta\lambda/\lambda$, of either 10 or 30%. The collimator entrance aperture and multi-detector can be a maximum distance of 40 m from the sample. For studies with lower resolution they are moved symmetrically towards the sample to a minimum distance of 1 m. Thus the accessible Q range is roughly

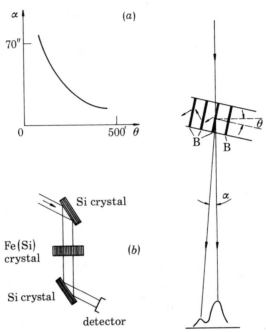

FIGURE 1. (a) Decrease of deflexion angle, α, with tilt angle, θ. (b) Illustration for α and θ; B are 90° Bloch walls (Schärpf 1977).

between 2×10^{-4} Å^{-1} and 0.3 Å^{-1}. The two-dimensional quadratic multi-detector with about 64×64 elements of size 1 cm^2 allows the simultaneous registration of the scattering for about 4000 Q values, which is of special value for anisotropic scattering patterns. The sensitivity may be characterized by the following statement: at D 11 a volume fraction of 10^{-4} of magnetic precipitates with a size of 300 Å is easily detectable in a sample of 1 cm^3. Additional counters (D 11B) in a circle around a special sample position can be used to get the diffuse scattering simultaneously. The D 11 instrument was shared between solid state physics, material scientists, chemists and biologists and thus a heavy overload resulted. Part of the burden at the I.L.L. has been taken over by a second instrument (D 17), constructed along the same principal lines but with slight modifications. There are similar or simpler instruments also at many other places (especially in west Europe), which together with D 11 and D 17 offer good possibilities in s.a.n.s. for the scientific community.

For magnetic scattering work it would be desirable to have polarized neutron beams or even the possibility of polarization analysis at the s.a.n.s. instruments. The former addition may be realized by a proper supermirror set-up. To get the addition of polarization analysis is certainly a more serious problem, if one would like to keep the multi-detector. A possible solution would be a nuclear polarization filter just behind the sample, if the filter could be produced in such a quality that no beam broadening or small-angle scattering arises from it. For more details on s.a.n.s. techniques the reader is referred to the recent review article by Schelten & Hendricks (1978).

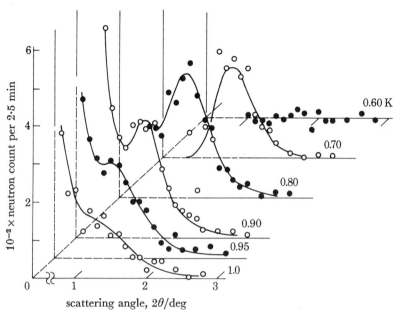

FIGURE 2. S.a.n.s. intensities of $ErRh_4B_4$ in the superconducting and low-temperature normal state (0.6 K) as a function of scattering angle 2θ and temperature (Moncton et al. 1979). The peak, corresponding to about 100 Å inhomogenity wavelength, is probably due to a helical magnetic structure.

A large variety of long-range magnetic fluctuations have been studied by neutron small-angle scattering: flux line lattices in type II superconducters, critical fluctuations in ferromagnets, superparamagnetic precipitates, precipitates in hard magnetic materials, dislocation strain induced fluctuations in cold-worked Ni and Fe near magnetic saturation, and magnetic fluctuations in the FeNi Invar alloy both as a function of concentration and temperature. Among the most recent studies the successful search for magnetic fluctuations in superconductors in the superconducting state should be mentioned especially (figure 2) (Moncton et al. 1979; Lynn & Glinka 1979).

5. MAGNETIC DIFFUSE SCATTERING

A typical magnetic scattering cross section for a binary alloy of about 1:1 ratio for the magnetic to non-magnetic component is 0.1–0.2 b/sr per atom†. This is well above the scattering cross section for coherent one-phonon scattering. Thus many studies for highly concentrated systems can be performed at normal two-axis diffractometer, and, if desirable, on triple-axis spectrometers and/or time-of-flight spectrometers. An essential drawback, however, is the unavoidable background due to nuclear incoherent elastic and nuclear elastic

† 1 barn (b) = 10^{-28} m².

disorder scattering. Depending on the system, there are favourable and extremely unfavourable systems. Also, in polarized beam experiments the nuclear scattering background reduces the statistical accuracy, and polarization analysis with a triple-axis spectrometer is at the very intensity limits. Nevertheless with 'normal' scattering techniques a considerable amount of work has been performed, especially by the Oak Ridge group.

What are the possible improvements? Diffuse scattering varies slowly with Q. This allows the use of large solid angles for the detector, multi-counter banks or multi-detectors can be installed, or the energy and angular resolution of the incident beam can be relaxed. Further, having a multi-detector reduction of inelastic background can in many cases be achieved by a low resolution time-of-flight technique. For instance, MAGS I at the FR2 in Karlsruhe is an instrument operating in this way. It is also no major problem to have an incident polarized beam in such an instrument. However, the only low-resolution spectrometer for diffuse scattering of thermal neutrons with polarization analysis is Longpol at Lucas Heights, at which iron transmission filters are used to guarantee large solid angles.

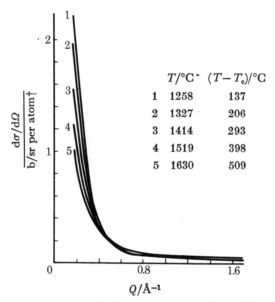

FIGURE 3. Diffuse scattering from magnetic clusters in cobalt below and above the melting temperature, $T_m = 1493$ °C (Rainer-Harbach 1979). † 1 barn (b) = 10^{-28} m².

An essential step towards higher intensity was achieved by the use of long-wavelength neutrons. Though the investigable Q-space is restricted for many magnetic systems, sufficient information can be obtained. The Harwell-glopper with a Be-filtered beam, a low resolution chopper and a detector bank was extremely successful in providing magnetic form factors for dilute alloys with a ferromagnetic matrix. The D7 spectrometer at the cold source of the H.F.R. in Grenoble is at present the instrument with highest sensitivity for diffuse magnetic scattering. It can be operated either with an unpolarized or with a polarized incident neutron beam, and has been used in both modes of operation successfully for many problems of magnetic disorder. A further development would be the addition of polarization analysis. A supermirror system for such an addition is under consideration now. As long as diffuse scattering studies can be performed with neutron wavelengths not shorter than 4 Å, instruments at cold sources are superior to those at thermal beams. To obtain optimum conditions for $\lambda > 4$ Å should be the major task in near future. Nevertheless, considered within a long-term policy, the situation also has to be improved for thermal neutrons; for polycrystalline materials and amorphous substances this is absolutely necessary.

Among the many investigations in diffuse magnetic scattering the most spectacular result, to my mind, was the analysis of the giant moment induced by iron in palladium. Also a very remarkable result was the series of measurements to determine the magnetic form of dilute substitutional atoms in ferromagnetic matrices also performed mainly by the Harwell group. Concentrated alloys have been studied to see whether the magnetic form factor remains even on dilution with non-magnetic atoms. An extreme case where this fails is given for NiCu alloys. By comparison of the nuclear with the magnetic disorder scattering cross section, a magnetic disturbance for the nearest neighbours of Cu (at 10% Cu) and magnetic disturbances up to fourth nearest neighbour shells for high Cu concentrations (40%) could be deduced. Nowadays, magnetic short-range order of spin-glass systems and systems at the ferromagnetic side of the percolation limit are of high interest. Another remarkable result obtained recently was the successful search for small magnetic clusters in molten Fe and Co (figure 3). By careful experiments with $\lambda = 0.7$ Å neutrons and proper corrections for spin dynamics for the magnetic clusters in Co 100 K above the melting temperature, a correlation length of $\zeta = 2.8$ Å was obtained, a value in good agreement with that expected by extrapolation from critical scattering (Rainer-Harbach 1979).

6. Inelastic magnetic diffuse scattering

It has been well known for many years that well above the Néel or Curie temperature, well localized moments have a time-dependent correlation function, $\langle m_i(0)\, m_i(t) \rangle \approx \exp(-t/\tau)$. Thus at high Q values we have quasielastic magnetic scattering described by a frequency spectrum proportional to $\omega_m/(\omega^2 + \omega_m^2)$, where $\omega_m = 1/\tau$. For small Q values, the neutron samples a larger region in real space and if there are many magnetic atoms within this region an excess magnetic moment diffuses more slowly out of this region than from a single site. ω_m becomes smaller; actually it becomes proportional to Q^2.

Recently the interest for spin diffusion has arisen again with all the studies of dilute magnetic moments in metals. For well behaved magnetic moments like Mn in Cu at high temperatures, or Fe in Au above the spin glass freezing temperature and also for many dilute 4f magnetic moments in metallic 4f compounds, Q-independent values of ω_m have been measured that are proportional to the temperature, the typical so-called Korringa behaviour. Depending on the coupling, $\hbar\omega_m(T)$ ranges between a small percentage and an appreciable amount of $k_B T$. For 3d magnetic impurities in non-magnetic metals like Cu and Au at very dilute concentrations, Kondo condensation may happen (e.g. Fe in Cu). Then $\hbar\omega_m$ tends to become constant with $T \to 0$, thus crossing the $k_B T$ line in an $\hbar\omega_m$–$k_B T$ diagram (figure 4a). The scattering cross section given above has to be multiplied by the detailed balance factor $\hbar\omega/(1-\exp(-\hbar\omega/k_B T))$ whereby with the proper normalization factor the energy-integrated scattering cross section remains approximately constant. At $k_B T < \hbar\omega_m$ this results in a diffuse magnetic scattering cross section, which is off-centred from $\omega = 0$ (figure 4b)! For a Kondo system, which requires very low concentrations, this behaviour is on the limits of observability. Nevertheless Loewenhaupt got some reasonable results with 480 Fe/10^6 in Cu! Much easier is the situation for intermediate valence systems, which show similar cross-over behaviour because the atomic concentration is here about 10–25%, and cross-over points occur at temperatures between 30 K ($Ce_2Cu_2Si_2$) and 300 K ($CeSn_3$) as far as investigations have been performed (Loewenhaupt & Holland-Moritz 1979). Such 'quasi-elastic' excitations seem also to exist for spin glasses below the freezing temperature, as observed recently for $Cu_{0.95}Mn_{0.05}$ with $\hbar\omega_m = 2.5$ meV, $k_B T = 4$ meV and $k_B T_f = 3$ meV (Scheuer et al. 1979). In interpreting the frequency spectrum with a single Lorentzian, caution seems to be necessary because of the restricted ω range normally applied. A recent experiment of Mezei & Murani (1979) with the neutron spin echo method demonstrated this clearly. With this technique, the time-dependent magnetic moment correlation function has been directly measured over a time scale of four orders of magnitude and it was evident that the data could not be interpreted by a single relaxation rate.

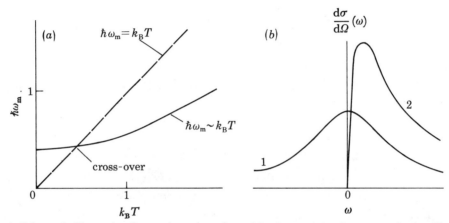

FIGURE 4. Schematic illustration for condensation of quasielastic scattering into an excitation-like, 'quasielastic' scattering for cross-over systems: (a) ω_m as a function of T; (b) frequency dependence of the scattering cross section for $\hbar\omega_m \ll k_B T$ (curve 1) and for $T = 0$ (curve 2).

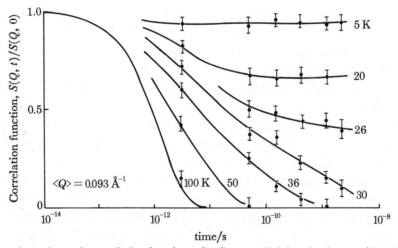

FIGURE 5. Time-dependent spin correlation functions of a Cu–5 at.% Mn spin glass as a function of temperature. Solid line, exponential spin correlation ($e^{-\gamma t}$) for $\gamma \approx \hbar\omega_m = 0.5$ meV. For other values of γ this line has only to be shifted to the left or the right.

References (Schmatz)

Badurek, G., Hammer, J. & Rauch, H. 1979 *J. Magn. magn. Mater.* **14**, 333.
Loewenhaupt, M. & Holland-Moreitz, E. 1979 *J. Magn. magn. Mater.* **14**, 227.
Lynn, J. W. & Glinka, C. J. 1979 *J. Magn. magn. Mater.* **14**, 179.
Mezei, F. & Murani, A. P. 1979 *J. Magn. magn. Mater.* **14**, 211.
Moncton, D. E., Shirane, G. & Thomlinson, W. 1979 *J. Magn. magn. Mater.* **14**, 172.
Rainer-Harbach, G. 1979 Ph.D. thesis, Universität Stuttgart.
Rekveldt, M. T. & van Schaik, F. J. 1979 *J. Magn. magn. Mater.* **14**, 325.
Schärpf, O. 1977 Habilitationsschrift at Technische Hochschule, Braunschweig.
Scheuer, H., Suck, J.-B. & Murani, A. P. 1979 *J. Magn. magn. Mater.* **14**, 241.
Schelten, J. & Hendricks, R. W. 1978 *J. appl. Crystallogr.* **11**, 297.
Schmatz, W. 1978 In *Disordered structures in neutron diffraction* (ed. H. Dachs), pp. 151–196. Berlin, Heidelberg and New York: Springer-Verlag.

The structure and dynamics of methane adsorbed on graphite

By G. Bomchil†, A. Hüller‡, T. Rayment§, S. J. Roser§,
M. V. Smalley§, R. K. Thomas§ and J. W. White†

† *Institut Max von Laue – Paul Langevin, Avenue des Martyrs 156X,
30842 Grenoble cedex, France*

‡ *Institut für Festkörperforschung, Jülich*

§ *Physical Chemistry Laboratory, South Parks Road, Oxford OX1 3QZ, U.K.*

The application of neutron scattering techniques to the study of methane physisorbed on graphite is described.

Two types of adsorbent have been used, graphitized carbon blacks of high and low surface area, and an exfoliated graphite.

Neutron diffraction has been used to determine the structure of the adsorbed layer at low temperatures, part of the phase diagram, and the distance of the methane molecule from the surface. Below a monolayer the methane molecules occupy a triangular lattice with a $\sqrt{3} \times \sqrt{3}$ structure in register with the underlying basal plane that forms the surface of graphite. Just above a monolayer, the lattice contracts out of register with the surface. The molecule–surface distance (carbon to surface) is found to be 3.30 ± 0.05 Å.

Incoherent neutron elastic scattering spectra give the frequencies of the vibrational modes of the adsorbed layer. The frequencies of both whole molecule displacements and torsional motions are found to be similar for directions perpendicular and parallel to the surface, at about 100 and 70 cm^{-1}, respectively.

Rotational tunnelling transitions have been observed in the range 0–200 µeV (0–1.6 cm^{-1}) corresponding to hindered rotation in a potential field of trigonal symmetry. The two types of barrier to rotation are estimated to be 150–200 cm^{-1} high.

All of the experimental parameters are compared with values calculated from atom–atom potentials by using different empirical parameters.

Introduction

The structure and dynamics of adsorbed monatomic and monomolecular layers are currently of interest (Dash 1975). This paper presents a progress report on recent neutron diffraction and spectroscopic measurements of the properties of methane adsorbed on the basal planes of graphite, and exemplifies the type of novel information on such systems that may be obtained by neutron methods (White *et al.* 1978).

The methane–graphite system follows a B.E.T. type II isotherm (Thomy & Duval 1970) characteristic of an adsorbate that completely 'wets' the surface of the adsorbent and therefore forms a monomolecular layer at suitable temperatures and pressures. Elsewhere we have described corresponding experiments on the ammonia–graphite system where the surface–molecule interaction is rather weak and which follows a B.E.T. type III isotherm (Bomchil *et al.* 1979 a; Gamlen *et al.* 1979). This is characteristic of an adsorbate that does not 'wet' the surface and that shows no tendency at all to form a monolayer. The experiments described in this paper are part of an attempt to find microscopic models for the two-dimensional physical and chemical behaviour of molecules in the monolayer next to the surface.

Materials, methods and thermodynamic properties

For the two types of experiment reported here, methane was used for the incoherent inelastic scattering experiments because of the high incoherent scattering cross section of protons, and deuteromethane (from Merck, Sharp and Dohme) for the diffraction experiments because of the high coherent scattering from deuterons. The relative contributions to the scattering of adsorbate and adsorbent have been summarized by Marlow et al. (1977, 1978) and White et al. (1978). Three different substrates were used, a high surface area (Vulcan III, 71 m² g⁻¹) and a low surface area (Sterling FT-G(2700), 11 m² g⁻¹) graphitized carbon black, both obtained from the National Physical Laboratory and described by Everett et al. (1974); and a partly orientated exfoliated graphite (Papyex, ca. 20 m² g⁻¹, mosaic spread ca. 30°) obtained from Le Carbone Lorraine (described by Coulomb et al. 1977).

The substrates, contained in welded aluminium 'cans' suitable for the neutron experiments, were outgassed for at least 12 h at 10^{-6} mmHg† and 350 °C immediately before the experiment. A background run of the scattering from the sample and container was always recorded before volumetric dosing of the surface with methane. This dosing was done slowly without any physical disturbance of the sample other than the changes in temperature required for complete annealing of the adsorbed layer. Annealing was carried out at temperatures where the methane molecules were known to have a liquid-like mobility (50–100 K). The coverages given in this work were calculated from the quoted surface areas of the adsorbents and a molecular area of 15.7 Å² for methane.

Preliminary neutron diffraction experiments from strongly diffracting samples were made with the Curran powder diffractometer at the Atomic Energy Research Establishment, Harwell, with an incident neutron wavelength of 2.63 Å‡. For weak reflexions and for studies of the temperature and coverage dependence of the peaks, the high flux diffractometer D1B at the Institut Laue–Langevin was used (Institut Laue–Langevin 1977). The incident flux and wavelength were 9×10^5 neutrons cm⁻² s⁻¹ and 2.4 Å. D1B has a multidetector, making it possible to record the whole pattern at once. Typically, with this instrument, it took 12 h to record a pattern from a 5 g sample at a coverage of 0.5 monolayers.

Neutron inelastic scattering experiments were also done at the Institut Laue–Langevin. Neutron energy loss spectra in the range 0–30 meV (0–240 cm⁻¹) at momentum transfers 1–5 Å⁻¹ were recorded on the rotating crystal time of flight spectrometer, IN4, at a resolution of $\Delta E/E$ of 5%. At lower energies, 0–5 meV (0–40 cm⁻¹), the cold neutron time of flight spectrometer, IN5, was used at a resolution of 20 µeV. The back scattering spectrometer, IN10, was used for energy transfers in the range 0–40 µeV (0–0.3 cm⁻¹) with a resolution of 1 µeV (0.008 cm⁻¹).

The methane–graphite system has been extensively studied by classical methods, and adsorption isotherms for many different carbon blacks have been determined over a wide range of temperature (Steele 1974; Thomy & Duval 1970). The isotherm shows the initial formation of a monolayer at low partial pressures of methane followed by multilayer deposition at higher pressures. For particularly homogeneous surfaces, discontinuities in the isotherm show the presence of different phases in the monolayer domain. These have been fully discussed by Thomy & Duval. Attempts to calculate the isosteric heat of adsorption from models of intermolecular forces have been made and are summarized by Steele.

† 1 mmHg ≈ 133.3 Pa. ‡ 1 Å = 10^{-10} m = 10^{-1} nm.

A fundamental difficulty in model calculations of isosteric heats is that there is neither a theoretical nor an empirical basis for the selection of parameters that characterize the molecule–surface potential. The problem is discussed at length by Steele (1974), and one of the aims of the present work is to produce a self-consistent potential that will describe accurately the structure and dynamics of adsorbed phases.

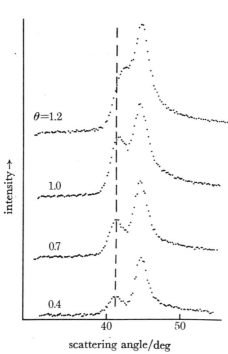

FIGURE 1. Diffraction patterns from methane-d_4 adsorbed at various coverages on Vulcan III at 20 K. The background patterns have been subtracted. The incident neutron wavelength was 2.4 Å. θ denotes coverage (see text).

STRUCTURAL STUDIES OF METHANE ADSORBED ON GRAPHITE

(a) The two-dimensional lattice

Figure 1 shows part of the neutron diffraction pattern from CD_4 adsorbed on Vulcan III at coverages (θ) between 0.4 and 1.2 monolayers and at a temperature of 20 K. Each pattern is the scattering remaining after subtraction of the background scattering from Vulcan III alone. At all coverages there are two peaks. The one at higher angle coincides with the (002) reflexion from the graphite basal planes and, as we shall discuss below, is associated with interference between the scattering from the adsorbed monolayer and from the basal planes. The lower angle peak at 41.4° ($\theta = 0.4$, $\lambda = 2.4$ Å) can only arise from diffraction by the absorbed layer of CD_4.

The asymmetry of the lower angle peak is characteristic of diffraction from a collection of randomly orientated two-dimensional lattices (Warren 1941). The low-angle edge is directly related to the methane–methane distance in the layer. For coverages between 0.4 and 0.9 this lattice parameter is 4.26 Å, consistent with a $\sqrt{3} \times \sqrt{3}$ hexagonal lattice in register with the basal planes of graphite that make up nearly all of the underlying surface. This agrees with our earlier results (Marlow et al. 1977, 1978) and with recent results of Vora et al. (1979) from methane on exfoliated graphite (Grafoil). Even at the lowest coverage, it is clear from the position and intensity of the peak that the molecules cluster into two-dimensional aggregates with the $\sqrt{3} \times \sqrt{3}$ structure rather than spreading uniformly over the surface to form a lattice gas. This shows that lateral attractive interactions between the methane molecules are relatively important in determining the overall structure. The best way to describe the contribution of these forces (as octupole–octupole or atom–atom interactions, for example) is still in doubt, despite recent theoretical calculations (O'Shea & Klein 1979; Maki & Nose 1979). We shall return to this question in the discussion below.

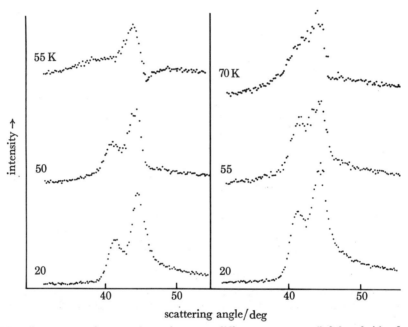

FIGURE 2. Diffraction patterns from methane-d_4 at two different coverages (left-hand side, $\theta = 0.7$; right-hand side, $\theta = 0.9$) and various temperatures on Vulcan III. The background patterns have been subtracted. The incident neutron wavelength was 2.4 Å.

When the coverage is increased above $\theta = 0.9$, the lattice parameter of the adsorbate decreases continuously, showing that the methane layer contracts out of register with the surface of the graphite. This behaviour is to be expected for two reasons. First, the van der Waals diameter of methane is 3.82 Å, markedly less than the methane–methane distance of 4.26 Å in the $\sqrt{3} \times \sqrt{3}$ structure. Thus at higher coverages the geometrical packing factor would be unreasonably low. Secondly, the interaction potential of graphite with an absorbed molecule does not vary much between different positions on the surface, so that little of the molecule–surface interaction energy is lost on going to a more compressed phase. There is a similar balance between these two opposing factors for nitrogen on graphite; this has been discussed by Steele (1977). At very high coverages (greater than two monolayers), the methane diffraction peak shifts until it is underneath the graphite (002) peak.

(b) The phase diagram of the adsorbate

The phase diagram for methane physisorbed on graphite can be constructed by following the temperature and coverage dependence of the diffraction peak associated with the two-dimensional lattice. Figure 2 shows the diffraction from CD_4 (background subtracted) between 20 and 70 K for two coverages.

At a coverage of 0.7, the main effect of an increase in temperature between 20 and 55 K is to broaden the diffraction peak. The considerable broadening between 50 and 55 K is associated with melting of the two-dimensional layer. Some melting has already occurred at 50 K and the range over which melting takes place is about 10 K, significantly greater than the range observed for methane on exfoliated graphite (Vora et al. 1979). A similar phenomenon has been observed for benzene on the two forms of graphite (Bomchil et al. 1979b). In the two-dimensional fluid phase at 55 K there is still plenty of intensity in the very broad diffraction peak, and its centre of gravity has shifted to lower angles, indicating an expansion of the layer. At a coverage of one monolayer, only a small fraction of the layer has melted at 55 K, and at 70 K the centre of gravity has not appreciably shifted from its low temperature position. In contrast, at $\theta = 0.4$, the melting point is lowered from about 55 K to about 50 K. We associate this behaviour with the effects of a smaller two-dimensional particle size, similar to the effects of particle size on the melting point of three-dimensional solids (Peppiatt & Sambles 1975). The two-dimensional particle size can be determined from a detailed analysis of the shape of the methane diffraction peak and we are currently attempting to use this information to obtain a quantitative interpretation of the melting point behaviour.

(c) The configuration of adsorbate molecules on the surface

The effect of an adsorbed layer on the diffraction pattern of the adsorbent may be to enhance or reduce the intensity of a given diffraction peak or even to change its shape to an extent that it appears to have shifted to a slightly different angle. The effect depends mainly upon the phase and amplitude of the scattering from the adsorbed layer relative to that from the series of crystal planes of which the surface is one. To a lesser extent it also depends on the size and shape of the adsorbent particles and the distribution and nature of defects in the appropriate set of crystal planes. The relative phase and amplitude of the scattering from the adsorbate depend on its configuration with respect to the surface, and a quantitative analysis of the interference can therefore yield valuable structural information. The sensitivity of the shape of the interference, or difference, peak to the methane–graphite distance is illustrated in figure 3 for the (004) reflexion from Vulcan III. The solid lines were calculated by using the simple model of Marlow et al. (1978) in which the shape of the difference peak is given by the product of the structure factor of the methane layer, taken as part of the one-dimensional lattice of the graphite basal planes, and the profile of the (004) reflexion in the absence of methane. The structure factor changes with angle in such a way that the difference peak is shifted by nearly 1° from the peak from the adsorbent alone. A similar effect has been observed by Larher et al. (1979) for adsorption on lead iodide.

The size of the cross interference effect seems to vary in the range ±10%, depending on both adsorbate and adsorbent. For methane on Vulcan III at 20 K and at a coverage of 0.7, the intensity of the (002) reflexion is enhanced by about 3%. To make accurate estimates of molecule–surface distances it is therefore essential to avoid any physical disturbance of the sample during adsorption so that the background subtraction may be as precise as possible. Careful corrections for both self shielding and multiple scattering must be made; ideally, a model should be fitted to the shape of both the difference peak and the reflexion from the adsorbent alone. An upper limit may be put on the experimental correction necessary for Vulcan by using the diffraction patterns measured for ammonia-d_3 on Vulcan (Gamlen *et al.* 1979). At low temperatures most of the ammonia is desorbed from the basal planes of graphite and there should therefore be negligible cross interference between the adsorbate and the graphite (00*l*) planes. Indeed, only small intensity changes are observed, less than 0.5%.

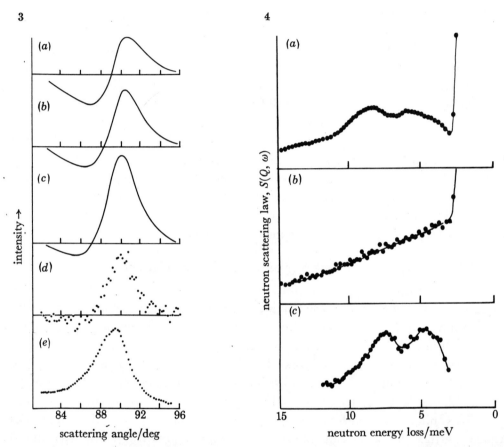

FIGURE 3. The interference effect of adsorbed CD_4 on the shape of the (004) reflexion of Vulcan III. (*a*), (*b*), (*c*) Profiles calculated for surface–molecule (carbon to carbon) distances of 3.25, 3.27 and 3.29 Å respectively; (*d*) the difference pattern observed at a coverage of one monolayer; (*e*) the profile observed for Vulcan III alone.

FIGURE 4. Vibrational spectra (neutron energy loss) of methane adsorbed on Vulcan III. (*a*) CH_4 at 10 K, coverage = 0.7; (*b*) CH_4 at 40 K, coverage = 0.7; (*c*) CD_4 at 10 K, coverage = 0.7. Incident neutron energy, 25 meV.

We have recently devised a quantitative model for describing the profiles of cross interference effects from layers adsorbed on carbon blacks. The model includes all of the factors mentioned above, as well as making allowance for thermal motion of the adsorbate. It will be described in full elsewhere. For methane at 20 K and at a coverage of 0.7 the molecule–surface distance (carbon to carbon) is found to be 3.3 ± 0.05 Å, as much as 10% shorter than predicted by many models of the molecule–surface potential (Steele 1974) though only slightly shorter than predicted by atom–atom potential calculations (see table 1). At higher temperatures and coverages, changes in the shape of the (002) difference peak occur which result from quite small changes in the molecule–surface distance. Thus the change in the (002) profile when the temperature changes from 50 to 55 K at a coverage of 0.4 (figure 2) is caused by a 10% increase in the molecule–surface distance when the two-dimensional layer melts.

THE DYNAMICS OF ADSORBED METHANE

The dynamics of the adsorbed layer can be studied by incoherent inelastic neutron scattering. Incoherent scattering is predominantly from motions involving displacement of hydrogen atoms and therefore the contribution from the adsorbate is much stronger than that from the carbon substrate. Also, there is no cross interference effect in incoherent scattering so that, provided that any coherent scattering can be eliminated, the spectrum resulting from motion of the absorbed layer can be obtained by straight subtraction of the spectrum of the adsorbent from that of adsorbent plus adsorbate. Another feature of incoherent scattering is that spectra may be measured at very high resolution (less than 1 µeV (0.008 cm^{-1})) down to low energy transfers, making it possible to measure accurately any very low energy transitions associated with rotational motion.

(a) Vibrational spectra

The partial differential cross section for incoherent one phonon scattering from a cubic lattice containing only one kind of atom is

$$\frac{\partial^2 \sigma_{\text{incoh}}}{\partial \Omega \, \partial E'} = \frac{k'}{k} \frac{N \sigma_{\text{incoh}}}{4\pi} \frac{Q^2 \langle u^2 \rangle}{2M} (n_s + \tfrac{1}{2} \pm \tfrac{1}{2}) \exp(-2W) \frac{Z(\omega)}{\omega},$$

where $Z(\omega)$ is the density of vibrational states, $\exp(-2W)$ the Debye–Waller factor, $\langle u^2 \rangle$ the mean square amplitude of vibration, M the mass of the vibrating atom, n_s the quantum number of mode s, $\boldsymbol{k}$ and $\boldsymbol{k}'$ the wavevectors (magnitudes k and k') of the ingoing and outgoing neutrons, Q, the momentum transfer, and σ_{incoh} the incoherent scattering cross section. The important features of the formula are twofold.

(i) If the incoherent scattering cross section of the atom is large, the spectrum is intense.

(ii) The intensity increases with both Q^2 and $\langle u^2 \rangle$ over the normally accessible range of Q. At higher values of Q the Debye–Waller factor will start to decrease rapidly and so reduce the intensity again. For anisotropic motion and an orientated sample, the formula becomes more complicated but the only important change is that the factor $Q^2 \langle u^2 \rangle$ is replaced by a term proportional to $\langle (\boldsymbol{Q} \cdot \boldsymbol{u})^2 \rangle$. The intensity of a band then depends on the relative orientation of $\boldsymbol{Q}$ and $\boldsymbol{u}$. This is an important factor when studying the spectra of adsorbates on orientated adsorbents such as Papyex. With large amplitude motions such as might be expected from physisorbed species, overtones and combinations (multiphonon processes) of fundamental vibrations may appear in the spectrum. Their intensities relative to the fundamentals will vary approximately as $(Q^2)^n$ where n is the number of quanta involved.

An adsorbed methane molecule has six degrees of freedom, two translations along axes parallel to the surface, $\nu_T(\|)$, one translation perpendicular to the surface, $\nu_T(\perp)$, one rotation about the perpendicular axis, $\nu_R(\perp)$, and two rotations about the parallel axes, $\nu_R(\|)$. In the limit of large barriers to both transition and rotation, these motions will become six vibrational modes which may be distinguished in several ways.

Deuteration will shift the translational modes by a relatively smaller amount than the rotational modes because the carbon atom moves in the former but not the latter. The modes will also be differently polarized. If Q is perpendicular to the surface, only $\nu_T(\perp)$ and $\nu_R(\parallel)$ will be observed. If Q is parallel to the surface, the spectrum will be dominated by $\nu_T(\parallel)$ and $\nu_R(\perp)$, while $\nu_R(\parallel)$ may also appear weakly. Finally some of the modes should be strongly dispersed. This will lead to a rather broad density of states giving a more diffuse spectrum whose maximum may also depend slightly on Q. Strongly dispersed modes will be those where the force field on a given methane molecule mainly originates from its interactions with neighbouring molecules. Since all theoretical calculations show that, for an isolated methane molecule on graphite, the barriers to translation across the surface or to rotation about an axis perpendicular to the surface are very small, we can expect either that the frequencies of $\nu_T(\parallel)$ and $\nu_R(\perp)$ will be near zero or that they will occur in the range up to about 100 cm^{-1} and show substantial dispersion. Taub et al. (1975) have measured the phonon spectrum of ^{36}Ar on graphite and shown that the strongly dispersed $\nu_T(\parallel)$ modes can be explained without invoking any surface molecule interaction.

Vibrational spectra of CH_4 on Vulcan were recorded over a range of coverage from 0.2 to 2 monolayers at temperatures from 10 to 70 K. Spectra of CD_4 on Vulcan and of CH_4 on Papyex were also measured. The spectra were recorded in neutron energy loss over a range 0–15 meV (0–120 cm^{-1}). Coherent scattering effects were eliminated by working at different wavelengths. The vibrational spectra of the adsorbate were obtained by subtraction of the background and, since there was only slight variation of the peak positions with Q, spectra recorded at different angles were added together to improve the statistics.

Figure 4a, b shows spectra from 0.7 monolayers of CH_4 on Vulcan at two different temperatures, 10 and 40 K. There are two pronounced peaks in the density of states at 10 K which completely disappear at 40 K even though the two-dimensional layer is still well below its melting point. The disappearance of the peaks implies that all the vibrations are strongly coupled to a diffusive motion at the higher temperature. Inelastic and quasielastic spectra at lower energy transfers (see below) show that, at 40 K, methane is undergoing rotational diffusion and it is therefore probable that all the vibrational modes are coupled to the rotation. In any case, it is clearly essential to measure the spectra of physisorbed species at as low a temperature as possible.

There are two strong peaks in the density of states at about 8.5 meV (70 cm^{-1}) and 6.2 meV (50 cm^{-1}) with a weaker peak at about 12.5 meV (100 cm^{-1}). Figure 4c shows the effects of deuteration (CD_4), again at a coverage of 0.7. The two peaks previously at 8.5 and 6.2 meV are now shifted to 7.6 (60 cm^{-1}) and 4.6 meV (38 cm^{-1}) giving isotope ratios of 1.15 and 1.35 respectively. The simplest interpretation would be that the higher energy peak is associated predominantly with translational modes while the lower energy peak is associated predominantly with rotation. The isotope shifts may, however, be misleading because of translation–rotation coupling and the overlap of widely dispersed modes.

Spectra from CH_4 on Papyex at a coverage of 0.7 are shown in figure 5. The general similarity of the spectra with Q perpendicular and Q parallel to the surface might suggest that the adsorbent is not very well orientated. The mosaic spread of the surface planes of Papyex is about 30°, and the material contains a proportion of randomly orientated crystallites. The contribution of vibrations with displacements in the direction of Q relative to those with displacements perpendicular to Q is

$$u^2 \cos^2 \theta / u^2 \sin^2 \theta,$$

so that even with Q misaligned by 30° the polarization is still strong. In measurements of the tunnelling transitions to be described below, the extent of polarization is observed to be almost complete so that it is reasonable to assume that it is also complete in the vibrational spectra. Since the spectra were obtained on a time of flight spectrometer there is also some variation in the direction of Q with energy. However, over the range of angles used for the spectra in figure 5 the direction of Q is within 10° of the specified direction up to energy transfers of about 12 meV.

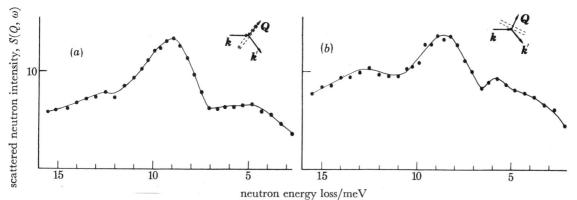

FIGURE 5. Vibrational spectra of methane adsorbed on exfoliated graphite (Papyex) at a coverage of 0.7 monolayers and a temperature of 10 K. (a) Q parallel to the surface; (b) Q perpendicular to the surface.

The main difference between the two spectra is in the relative intensity of the bands, the band at 12.5 meV being markedly stronger when Q is perpendicular to the surface. There are also small differences in energy. For example, the central peak is at 8.5 meV when Q is perpendicular but at 9 meV when Q is parallel. When Q is perpendicular, we expect to observe only the translational motion perpendicular to the surface ($\nu_T(\perp)$) and the two rotations about axes parallel to the surface ($\nu_R(\|)$). On the basis of intensity and polarization we assign the peak at 8.5 meV to $\nu_R(\|)$ and the peak at about 13 meV to $\nu_T(\perp)$. When Q is parallel to the surface, we expect to see two translational modes ($\nu_T(\|)$) and one rotational mode ($\nu_R(\perp)$) both strongly dispersed. Again, rotational modes are expected to be the most intense and we therefore assign the peak at 9 meV to $\nu_R(\perp)$. Taub et al. (1975) have calculated the dispersion curves for $\nu_T(\|)$ for two-dimensional argon on a triangular lattice by using a Lennard–Jones interaction potential. There are two main peaks in the density of states at 6 and 3.5 meV. Methane has the same lattice symmetry and its Lennard–Jones parameters are sufficiently similar to those of argon that the only difference between the two will be their different masses. On this basis the peaks in the translational density of states of methane would be at 9 and 5.3 meV. Thus the peak at 9 meV in figure 5a may also be partly due to translational motion parallel to the surface, which would be more consistent with the isotope effect. The peak at about 5 meV may then be the second maximum in the translational density of states. An analysis of the tunnelling spectrum presented below requires $\nu_R(\perp)$ to be close in frequency to $\nu_R(\|)$ consistent with the assignment given.

The final assignments of the vibrations of adsorbed methane are given in table 1, and $\nu_T(\perp)$ is compared with calculated frequencies in table 3. The similarity of the energies of the three rotational modes indicates that an adsorbed methane molecule is in a fairly isotropic potential field. This is discussed further below.

TABLE 1. WAVENUMBERS OF MAXIMA IN THE DENSITY OF STATES OF METHANE ADSORBED ON GRAPHITE (PAPYEX)

wavenumber/cm^{-1}	assignment	description
35–50	$\nu_T(\|)$	translational motion on plane of surface, some contribution from rotation
47±2	$\nu_R(\|)$	torsion about axes parallel to surface
69±3	$\nu_R(\|)$	rotation of molecule about axes parallel to surface
73±3	$\nu_R(\perp)$	rotational motion in plane of surface
100–110	$\nu_T(\perp)$	displacement of whole molecule in direction perpendicular to surface

(b) *Rotational motion*

For a free methane molecule the $J = 0 \to 1$ rotational transition requires an energy of 1.25 meV (10 cm^{-1}). For coverages less than one monolayer, no free rotational transition was observed in the incoherent scattering spectrum. This is as expected because the surface potential must be quite strong to give rise to the energies of torsional oscillations given in table 1. However, for a bilayer of methane on Vulcan a distinct excitation was observed at 720 µeV, indicating a slightly hindered rotation of methane molecules in the second layer (Newbery *et al.* 1978). The energy of the transition is close to that observed for methane molecules isolated in rare gas matrices (Kataoka *et al.* 1978).

(c) *Rotational tunnelling*

For coverages of methane less than about a monolayer, sharp transitions between 0 and 200 µeV (1.6 cm^{-1}) are observed in the high-resolution spectrum of methane adsorbed on Vulcan. These have been attributed to rotational tunnelling of the methane molecule in the surface potential field (Newbery *et al.* 1978). Figure 6 shows the spectra for CH$_4$ on Vulcan at a coverage of 0.7 and at a number of temperatures. They are similar in appearance to the tunnelling spectra of solid methane (Press & Kollmar 1975) but with transition energies lower by about 25%. This similarity suggests an unexpected isotropy of the crystal potential around an adsorbed methane molecule.

To resolve any finer structure in the rotational tunnelling spectrum, measurements were made with Papyex as adsorbent and with the momentum transfer, Q, parallel and perpendicular to the surface planes. The two spectra and their assignments are shown in figure 7. Each of the transitions previously observed (figure 6) is now split into two and most of them are strongly polarized. With Q parallel to the surface the central elastic peak is also broadened, suggesting that there is a further low-energy transition, not observed owing to lack of resolution at the IN5 spectrometer. This transition has now been observed directly at 17 µeV (0.14 cm^{-1}) by using the 1 µeV resolution of IN10 and is shown in figure 8 *b*. Its observation allows the complete energy level diagram to be drawn, summarizing the five nuclear spin allowed transitions. This is shown in figure 8 *a* and the energies of the transitions are given in table 2.

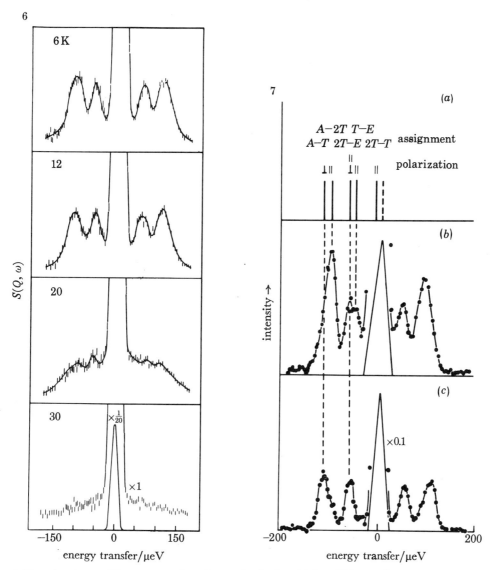

Figure 6. Rotational tunnelling spectra of methane adsorbed on Vulcan III at various temperatures. Coverage, 0.7 monolayers.

Figure 7. Rotational tunnelling spectra of methane adsorbed on exfoliated graphite (Papyex). (a) Assignment and polarization of transitions; (b) Q parallel to the surface; (c) Q perpendicular to the surface. Coverage, 0.7; temperature, 4 K.

Table 2. Energies of rotational tunnelling transitions of adsorbed methane

energy/μeV	assignment
17 ± 1	$2T \to T$
39 ± 2	$T \to E$
56 ± 2	$2T \to E$
94 ± 2	$A \to 2T$
112 ± 2	$A \to T$

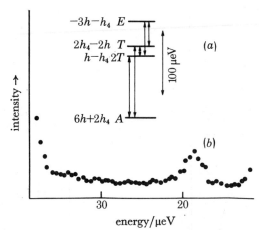

FIGURE 8. High-resolution spectrum of the rotational tunnelling spectrum of methane on exfoliated graphite (Papyex). Coverage, 0.7 layers; temperature, 4 K; Q parallel to the surface. The energy levels are given in (a) in terms of the two parameters h and h_4 (see text).

The energy level pattern shown in figure 8a is characteristic of a tetrahedral molecule tunnelling in a potential field with trigonal symmetry but which is not far from tetrahedral. In a tetrahedral field the two central levels coalesce and the energy of the $A \to T$ transition should then be approximately twice that of the $T \to E$ transition, as has been observed for phase II of solid CH_4 (Press & Kollmar 1975) and as is observed here for the centre of gravity of the two transitions for adsorbed methane. The earlier conclusion of Newbery et al. (1978), that the adsorbed methane on Vulcan was undergoing rotational diffusion about an axis perpendicular to the surface, was based on the presence of an envelope of scattering underneath the tunnelling peaks which resembled quasielastic scattering. This envelope is not present in the Papyex spectra, suggesting that it has its origin in methane molecules adsorbed at heterogeneities on the Vulcan or at defects in the two-dimensional layer.

The general procedure for calculating the rotational tunnelling levels has been described by Hüller & Kroll (1975). The potential energy is expanded as a series of cubic rotator functions whose allowed coefficients are determined by molecule and site symmetry. As in rotational tunnelling about one axis, the series rapidly converges and it is necessary to include only the lowest terms in the expansion. If the barrier to rotation is not too low, 'pocket state' wavefunctions form an adequate basis set. These have the properties that they are large in the region of a potential minimum but small elsewhere, they have an adjustable parameter so that a variational treatment may be used to minimize the final energies of the tunnelling levels, and they are readily adapted to the symmetry of the problem. The pocket states are not exact wavefunctions because of the finite probability of tunnelling between potential minima, and so the Hamiltonian is not diagonal. By using the pocket state wavefunctions as a basis, the matrix elements may be evaluated and the eigenvalues minimized with respect to the variational parameter. Because they are small, these off-diagonal elements in the Hamiltonian determine the tunnelling splitting directly. The energy levels may therefore be expressed in terms of these elements without explicitly evaluating them.

Hüller (1977) has derived a Hamiltonian for CH_4 in a tetrahedral field which may easily be adapted to the present problem. There are three matrix elements that might be important, H, h and h_4, which are respectively the matrix elements for 180° rotation about twofold axes in CH_4, 120° rotation about one of the threefold axes pointing towards the surface, and 120° rotation about the threefold axis perpendicular to the surface. H is likely to be much smaller than h or h_4 because the overlap between pocket states 180° apart will be much less than between those that are 120° apart. Also, the ratio of the energies of the $A \to T$ and $T \to E$ transitions becomes exactly two, as observed, only when H is insignificant compared with h and h_4. Neglecting H, the relative spacing of the levels in terms of h and h_4 is given in figure 8a.

The splitting of the T levels is proportional to the difference between h and h_4, but it is not possible to determine the sign of the splitting from the energies of the tunnelling transitions alone. However, the ratio of the intensities of the A–T transitions with Q parallel and perpendicular to the surface is about two, suggesting on a simple argument of degeneracy that the former is the $A \to 2T$ transition. This completes the assignment of the transitions given in table 2, from which h and h_4 are found to be -14 and -8 μeV (0.11 and 0.06 cm^{-1}) respectively.

Since $|h_4|$ is smaller than $|h|$, the overlap between the pocket state wavefunctions must be smaller for rotation about an axis perpendicular to the surface than about the three equivalent C–H bonds pointing towards the surface. The barrier to rotation about the perpendicular axis must therefore be the greater one. We are currently making a full quantitative analysis of h and h_4 to obtain accurate values of the two barriers. At present, we can only give an approximate estimate of 150–200 cm^{-1} for both barrier heights with one about 25% greater than the other. The similarity of the two barrier heights means that the two torsional frequencies, $\nu_R(\perp)$ and $\nu_R(\|)$, must be similar in frequency.

Discussion

Table 3 summarizes the experimental parameters for methane on graphite as determined by neutron scattering. This is the first time that such a complete set of data has been assembled for a physisorbed molecule. Also included for comparison are the corresponding quantities calculated by using atom–atom potentials with different sets of empirical parameters. The calculation is based on a potential of the form

$$V = \tfrac{1}{2} \sum_i V(r_i) = -A/r_i^6 + B \exp(-Cr_i),$$

where A, B and C are constants appropriate to pairs of atoms (Williams 1967). Interactions with atoms in the first layer of the surface out to a distance of 18 Å from the methane molecule have been included. Terms arising from molecule–surface and molecule–molecule interactions are given separately and values are given for the methane molecule in two possible positions on the surface. The first and second differentials of the potential were used to obtain the equilibrium surface–molecule distance and the vibrational force constant respectively.

Some of the empirical parameter sets account quite well for the isosteric heat of adsorption and the distance of methane from the surface but they all grossly overestimate the changes in potential energy for any kind of displacement from the equilibrium position. The calculated values of the vibrational frequencies and barriers to rotation are therefore all too high. Despite the obvious deficiencies of the atom–atom calculations, they do allow some rationalization of the structural information derived from the diffraction data.

The methane–methane interaction energy calculated for the tripod configuration is a large fraction of the isosteric heat of adsorption. This interaction also provides the cohesion necessary for the formation of aggregates of methane molecules as observed at coverages down to 0.4

TABLE 3. COMPARISON OF ATOM–ATOM POTENTIAL CALCULATIONS AND EXPERIMENTAL MEASUREMENTS FOR THE $\sqrt{3} \times \sqrt{3}$ PHASE OF METHANE ADSORBED ON GRAPHITE

source of empirical parameters	binding energy of isolated molecule kJ mol^{-1}	methane–methane interaction energy kJ mol^{-1} [cm^{-1}]	isosteric heat of adsorption at 130 K and zero coverage kJ mol^{-1} [cm^{-1}]	distance from surface (C–C) nm	vibration frequency perpendicular to surface E cm^{-1}	barrier to rotation about perpendicular axis/cm^{-1} methane–surface contribution	barrier to rotation about perpendicular axis/cm^{-1} methane–methane contribution	barrier to rotation about perpendicular axis/cm^{-1} total	barrier to rotation about C–H bond/cm^{-1}
Kitaigorodskii†	17.6 (17.3)	6.6 [505]	14.9 [1140]	0.325 (0.327)	102	57 (2)	582	639 (580)	614 (367)
Taddei (A)‡	17.5 (17.4)	5.2 [400]	14.8 [1130]	0.333 (0.333)	98	40 (2)	359	399 (357)	447 (288)
Taddei (B)‡	15.7 (15.8)	3.5 [270]	13.0 [990]	0.335 (0.335)	93	26 (1)	143	169 (142)	291 (196)
Williams no. 4§	16.6 (16.8)	3.8 [290]	13.9 [1060]	0.332 (0.331)	86	20 (1)	261	281 (260)	207 (173)
Williams‖	14.9 (15.0)	3.5 [270]	12.2 [930]	0.333 (0.332)	93	19 (1)	207	226 (206)	208 (173)
experiment	—	—	12.3¶	0.330±0.005	100±5	—	—	150–200	150–200

References: †, Kitaigorodskii (1973); ‡, Taddei et al. (1977); §, Williams (1967); ‖, Williams (1970); ¶, Kiselev & Poshkus (1976).

The barrier to rotation about C–H bond refers to one of the three equivalent C–H bonds pointing towards the surface. Values in parentheses are calculated for methane over the centre of a carbon hexagon, the others are for methane directly over a carbon atom. The isosteric heat is obtained from the binding energy by adding $\frac{5}{2}RT$ (three rotational, two translational, and one vibrational degree of freedom, $q_{\rm st} = -\Delta U_0^0 - \frac{5}{2}RT$). The calculated vibration frequencies are those appropriate to a Morse potential.

monolayers. The tunnelling spectrum observed at 0.2 monolayers is similar to that observed at higher coverages showing that the aggregates persist to even lower coverages.

At a coverage of 0.9 monolayers the $\sqrt{3} \times \sqrt{3}$ structure of adsorbed methane changes to a more compressed phase. The lattice parameter decreases continuously over a range of coverage. Our results with Vulcan III as adsorbent agree here both with the results of Vora *et al.* (1979) for exfoliated graphite and with the theoretical predictions of Bak *et al.* (1979) for such a phase change in a two-dimensional lattice. A major factor allowing this phase change is the small variation of the surface–molecule interaction energy across the surface (column 1 of table 3). Indeed, the surface is so smooth that it is not possible to predict whether the equilibrium position of the methane in its low coverage phase is above the centre of a carbon hexagon or directly above one of the carbon atoms on the surface. Further studies of the intensity changes of the graphite (10) reflexion on adsorption are needed to resolve this point. Carlos & Cole (1979) have suggested that anisotropic pair potentials give a better description of the atom–graphite interaction and, since this potential enhances the difference in energy between the two sites, application of such anisotropy to the atom–atom calculations might distinguish more clearly between the two possible configurations. Pursuit of this potential is at present the most obvious single improvement that could be made to available potentials.

The temperature dependence of the tunnelling splitting presents new features compared with the behaviour of molecular crystals, where a distinct shift in energy as well as broadening often occurs. At present there is no known mechanism for the behaviour observed.

References (Bomchil *et al.*)

Bak, P., Mukanal, D., Villain, J. & Westowska, K. 1979 *Phys. Rev.* B **19**, 1610.
Bomchil, G., Harris, N. M., Leslie, M., Tabony, J., White, J. W., Gamlen, P. H., Thomas, R. K. & Trewern, T. D. 1979a *J. chem. Soc. Faraday Trans.* I **75**, 1535.
Bomchil, G., Meehan, P., Rayment, T., Thomas, R. K. & White, J. W. 1979b *Surf. Sci.* (In the press.)
Carlos, W. E. & Cole, M. W. 1979 *Phys. Rev. Lett.* **43**, 697.
Coulomb, J. P., Bienfait, M. & Thorel, P. 1977 *J. Phys., Paris* C **38** (4), 31.
Dash, J. G. 1975 *Films on solid surfaces*. New York: Academic Press.
Everett, D. H., Parfitt, G. D., Sing, K. S. W. & Wilson, R. 1974 *J. appl. Chem. Biotechnol.* **24**, 199.
Gamlen, P. H., Thomas, R. K., Trewern, T. D., Bomchil, G., Harris, N. M., Leslie, M., Tabony, J. & White, J. W. 1979 *J. chem. Soc. Faraday Trans.* I **75**, 1542, 1553.
Hüller, A. 1977 *Phys. Rev.* **16**, 1844.
Hüller, A. & Kroll, D. M. 1975 *J. chem. Phys.* **63**, 4495.
Kataoka, Y., Press, W., Buchenau, U. & Spitzer, H. 1978 In *Inelastic neutron scattering 1977*. Vienna: International Atomic Energy Agency.
Kiselev, A. V. & Poshkus, D. P. 1976 *J. chem. Soc. Faraday Trans.* II **72**, 950.
Kitaigorodskii, A. E. 1973 *Molecular crystals*. New York: Academic Press.
Larher, Y., Thorel, P., Gilquin, B., Croset, B. & Marti, C. 1979 *Surf. Sci.* **85**, 94.
Maki, K. & Nose, S. 1979 *J. chem. Phys.* **71**, 1392.
Marlow, I., Thomas, R. K., Trewern, T. & White, J. W. 1977 *J. Phys., Paris* C **38** (4), 19.
Marlow, I., Thomas, R. K., Trewern, T. & White, J. W. 1978 In *Inelastic neutron scattering 1977*. Vienna: International Atomic Energy Agency.
Newbery, M. W., Rayment, T., Smalley, M. V., Thomas, R. K. & White, J. W. 1978 *Chem. Phys. Lett.* **59**, 461.
O'Shea, S. F. & Klein, M. 1979 *J. chem. Phys.* **71**, 2399.
Peppiatt, S. J. & Sambles, J. R. 1975 *Proc. R. Soc. Lond.* A **345**, 387.
Press, W. & Kollmar, A. 1975 *Solid State Commun.* **17**, 405.
Steele, W. A. 1974 *The interaction of gases with solid surfaces*. Oxford: Pergamon Press.
Steele, W. A. 1977 *J. Phys., Paris* C **38** (4), 61.
Taddei, G., Righini, R. & Manzetti, P. 1977 *Acta crystallogr.* A **33**, 626.
Taub, H., Passell, L., Kjems, J. K., Carneiro, K., McTague, J. P. & Dash, J. G. 1975 *Phys. Rev. Lett.* **34**, 654.
Thomy, A. & Duval, X. 1970 *J. Chim. phys.* **67**, 1101.

Vora, P., Sinha, S. K. & Crawford, R. K. 1979 *Phys. Rev. Lett.* (In the press.)
Warren, B. E. 1941 *Phys. Rev.* **59**, 693.
White, J. W., Thomas, R. K., Trewern, T., Marlow, I., Bomchil, G. 1978 *Surf. Sci.* **76**, 13.
Williams, D. E. 1967 *J. chem. Phys.* **47**, 4680.
Williams, D. E. 1970 *Am. crystallogr. Ass.* **6**, 21.

Discussion

A. D. BUCKINGHAM, F.R.S. (*University Chemical Laboratory, Cambridge, U.K.*). Could Dr White clarify the nature of the torsional vibrations of the adsorbed methane molecules? What is the source of the potential barrier?

Secondly, would Dr White relate the potential energy deduced to other experimental results, for example the enthalpy of adsorption? Is there evidence for non-additive potentials?

J. W. WHITE. Table 3 in the paper provides a number of the answers to Professor Buckingham's questions. The results are calculations for the $\sqrt{3} \times \sqrt{3}$ registered phase at low temperatures.

From the first two columns one can see that both graphite–methane and methane–methane interactions contribute to the hindering potential for molecular motion. Assuming a tripod configuration for the adsorbed molecule, the torsional vibrations occur about a C3 axis perpendicular to the surface and about another pair of axes, which may be chosen parallel to the surface.

Since there is coupling between neighbouring molecules, the torsional modes are almost certainly dispersed in the two-dimensional Brillouin zone. What we observe are the density of states maxima. The tunnelling results give a precise method of testing models for the potential barrier to rotation.

It can be seen that for either set of axes all of the commonly used 6-exp, atom–atom potential parameter sets give barriers to rotation that are too high. The calculated enthalpies of adsorption are also somewhat high.

I presume that by non-additive potentials Professor Buckingham means non-additivity of the molecule–surface and molecule–molecule parts to give, for example, isosteric heat of adsorption. The calculated values certainly do not add up to those found experimentally. As commented on in the paper, we are not sure whether to attribute this to anisotropic polarisation effects in the potential for adsorption on graphite or to real non-additive effects (Axelrod–Teller effects, etc.).

Neutron diffraction, isotopic substitution and the structure of aqueous solutions

By J. E. Enderby

*H. H. Wills Physics Laboratory, University of Bristol,
Royal Fort, Tyndall Avenue, Bristol BS8 1TL, U.K.*

The structural properties of liquids that contain more than one atomic species are difficult to unravel. The essential reason for this is that conventional diffraction experiments measure a rather coarse average of all the $\frac{1}{2}\nu(\nu+1)$ partial structure factors that characterize a liquid containing ν species.

This paper describes the way that neutron diffraction experiments carried out on samples that are identical in all respects, except that the isotope of one or more of the species has been changed, can overcome this problem. It is shown that the systematic use of isotopes, by virtue of the dependence of the neutron scattering amplitude on isotope, enables partial structure factors to be extracted directly from diffraction data. A detailed account of the method applied to a range of aqueous solutions of electrolytes under various experimental conditions is given.

In particular, *first-order difference* experiments yield information on ion–water conformations. Data are now available for the cations Ni^{2+}, Ca^{2+} and Li^+ and these results will be discussed in detail. The data for Ni^{2+} are of particular interest because they show that the substantial angle of tilt between the plane of the water molecule and the Ni–O axis gradually disappears as the concentration is reduced.

The only anion studied so far is Cl^- but the experiments have been carried out for a wide range of counter ions. We have shown that for such different electrolytes as $CaCl_2$, NaCl and LiCl, the nature of the hydration around the Cl^- ion is essentially the same.

Finally, the method of *second-order difference* yields directly ion–ion correlations. The experiments described include Cl–Cl structure factors in NaCl and $NiCl_2$ solutions, and the Ni–Ni structure factor in $NiCl_2$ solutions. Comparisons made with theoretical predictions based on the primitive model of electrolytes show that in certain cases, the molecular nature of the water is a crucial factor in determining ion–ion correlations. In other cases, the primitive model contains most of the essential physics.

1. Introduction

In the extensive literature devoted to the structural problem of aqueous electrolyte solutions, two distinct themes have emerged. The first of these is to do with the coordination, orientation and conformation of the water–ion subsystem. This aspect of the solution problem, usually referred to as hydration, has been the subject of an enormous amount of experimental effort. It is however, fair comment that until recently important questions like, for example, the detailed dependence on concentration of the conformation of the ion–water system or the sensitivity of hydration to the nature of the counter-ion were largely unanswered. The theoretical methods available to discuss hydration are extensive and range from the techniques of coordination chemistry through quantum mechanical cluster calculations and finally to computer simulations based on a model water potential. What is badly needed is a sharp experimental test of these theories; we believe that the method of neutron diffraction on isotopically enriched samples described below meets that need.

554   J. E. ENDERBY

The second general theme concerns the correlations between ions. Here we can discern two important sub-themes. In the 'primitive' view of electrolytes, the molecular nature of the water is neglected and the problem becomes the classic one in liquid state physics: given a known (and relatively simple) interaction, what structure will result? At the other extreme, essentially chemical notions like complexing, ion-pairing and stable species are used. The validity of both approaches can be tested in terms of *ion–ion* correlation functions and once again these are accessible through the isotopic substitution method. In this paper I shall review what has been achieved so far and what the future prospects are for this subject.

2. THE METHOD

We have shown that the neutron 'first order' difference method (Soper et al. 1977; Neilson & Enderby 1978; Enderby & Neilson 1980) allows one to gain direct information about the detailed arrangement of the water molecules around the ions in aqueous solutions. Let us consider a salt MX_n (M = metal; X = halide) dissolved in heavy water, D_2O, and let c represent the atomic fraction of M. The quantity that is central to the method is the algebraic difference in the absolute differential scattering cross section as a function of the scattering vector k from two samples that are identical in all respects except that the isotopic state of M (or X) has been changed; this quantity, denoted $\Delta_M(k)$ or $\Delta_X(k)$ is the sum of four partial structure factors $S_{\alpha\beta}(k)$ weighted in such a way that only those relating to ion–water correlations are significant. Explicitly:

$$\Delta_M(k) = A_1(S_{MO}(k)-1) + B_1(S_{MD}(k)-1) + C_1(S_{MX}(k)-1) + D_1(S_{MM}(k)-1) \tag{1}$$

and
$$\Delta_X(k) = A_2(S_{XO}(k)-1) + B_2(S_{XD}(k)-1) + C_2(S_{MX}(k)-1) + D_2(S_{XX}(k)-1), \tag{2}$$

where
$$A_1 = \tfrac{2}{3}c(1-c-nc)f_O(f_M-f'_M); \quad A_2 = \tfrac{2}{3}nc(1-c-nc)f_O(f_X-f'_X);$$
$$B_1 = \tfrac{4}{3}c(1-c-nc)f_D(f_M-f'_M); \quad B_2 = \tfrac{4}{3}nc(1-c-nc)f_D(f_X-f'_X);$$
$$C_1 = 2nc^2 f_X(f_M-f'_M); \quad C_2 = 2nc^2 f_M(f_X-f'_X); \quad \text{and}$$
$$D_1 = c^2(f_M^2-(f'_M)^2); \quad D_2 = n^2c^2(f_X^2-(f'_X)^2).$$

Here f_O and f_D are the neutron coherent scattering amplitudes for oxygen and deuterium and f_M, f'_M, f_X and f'_X are the scattering amplitudes for the isotopic states used in producing the salt MX_n.

The properties of $\Delta(k)$ have been discussed in detail elsewhere (Soper et al. 1977) and need not be enlarged on here. The crucial property, apart from the fact that $A, B > C, D$, is that the high k Placzek distortions in the differential scattering cross section from aqueous solutions are essentially eliminated so that a weighted distribution function $G(r)$ can be determined from

$$G(r) = \frac{1}{2\pi^2 \rho r} \int \Delta(k)\, k \sin(kr)\, dk, \tag{3}$$

where ρ is the total number density, by standard numerical quadrature. In terms of the partial radial distribution functions, $g_{\alpha\beta}(r)$, it follows at once that

$$G_M(r) = A_1(g_{MO}-1) + B_1(g_{MD}-1) + C_1(g_{MX}-1) + D_1(g_{MM}-1) \tag{4}$$

and
$$G_X(r) = A_2(g_{XO}-1) + B_2(g_{XD}-1) + C_2(g_{MX}-1) + D_2(g_{XX}-1). \tag{5}$$

[74]

Since A and B are much greater than C and D the method yields a high resolution measurement of an appropriate combination of g_{MO} and g_{MD} or g_{XO} and g_{XD}.

The 'second order' difference method (Howe et al. 1974; Enderby & Neilson 1980) allows one to gain direct information about ion–ion correlations. Let us again consider a salt MX_n dissolved in D_2O and let $\Delta_{M_1}(k)$ and $\Delta_{M_2}(k)$ represent the two first-order differences for three solutions with M in the isotopic state M, 'M and "M. Similarly let $\Delta_{X_1}(k)$ and $\Delta_{X_2}(k)$ be the corresponding quantities for isotopic substitutions of the anion X. It follows from equations (1) and (2) that

$$S_{MM}(k) = \frac{\Delta_{M_1}(k)}{A_M^{(2)}} - \frac{\Delta_{M_2}(k)}{B_M^{(2)}} + 1; \tag{6}$$

$$S_{XX}(k) = \frac{\Delta_{X_1}(k)}{A_X^{(2)}} - \frac{\Delta_{X_2}(k)}{B_X^{(2)}} + 1; \tag{7}$$

with coefficients $A_M \ldots B_X$ given by

$$A_M^{(2)} = c^2(f_M - f_M')(f_M' - f_M''); \quad B_M^{(2)} = c^2(f_M - f_M'')(f_M' - f_M'');$$

$$A_X^{(2)} = n^2c^2(f_X - f_X')(f_X' - f_X''); \quad B_X^{(2)} = n^2c^2(f_X - f_X'')(f_X' - f_X'').$$

To obtain the cross term S_{MX}, we must use four samples whose isotopic state can be represented by

$$MX_{n'}, \; 'MX_{n'}, \; M'X_{n'}, \; 'M'X_{n}.$$

Let Δ_M^X represent the algebraic difference in intensity between the scattering from the first and the second samples, and $\Delta_M'^X$ the difference between the third and the fourth sample. Straightforward manipulation yields

$$S_{MX}(k) = \frac{\Delta_M^X - \Delta_M'^X}{2nc^2(f_M - f_M')(f_X - f_X')} + 1. \tag{8}$$

In real space, the three ion–ion radial distribution functions can be obtained through

$$g_{\alpha\beta}(r) = \frac{1}{2\pi^2 \rho r} \int [S_{\alpha\beta}(k) - 1] \, k \sin(kr) \, dk, \tag{9}$$

by analogy with (3) with $\alpha, \beta = M$ or X.

3. Experimental results

3.1. *The hydration of cations*

In the research programme carried out so far, three cations have been investigated. Salts of $NiCl_2$ (isotopically changing the Ni), $CaCl_2$ (isotopically changing the Ca) and LiCl (isotopically changing the Li) were dissolved in D_2O and the concentrations and coefficients $A_1, B_1, C_1,$ and D_1 are given in table 1. The statement made in §2 that C_1 and D_1 are small compared with A_1 and B_1 is evidently true. We now consider each of these cations in turn.

3.1.1. Ni^{2+}

Our evidence on Ni^{2+} hydration derives from the detailed study of $NiCl_2$ solutions made over a wide range of concentrations by Neilson & Enderby (1978). An example of $\Delta_{Ni}(k)$ and of $G_{Ni}(r)$ is shown in figures 1 and 2. It is particularly significant that at $r \approx 3.0$ Å†, $G_{Ni}(r)$

† $1 \text{ Å} = 10^{-10} \text{ m} = 10^{-1} \text{ nm}$.

556   J. E. ENDERBY

close to the sum $-(A_1+B_1+C_1+D_1)$ implying (equation (4)) that g_{NiO} and g_{NiD} are both small at this value of r; this reflects the stability of the first hydration shell. The two peaks located at 2.07 and 2.67 Å can be identified with Ni–O and Ni–D correlations respectively, on the grounds that the ratio of the areas beneath them, when weighted by r^2, is almost exactly $A_1:B_1$. An integral over $4\pi r^2 G_{Ni}(r)$ for $1.8 < r < 3.0$ Å yields the number of water molecules in the first coordination shell, the so-called *hydration number*. We are therefore able for the first time to investigate the concentration dependence of the conformation of Ni–D$_2$O provided we know the geometry of the water molecule.

For all plausible values of the bond length and angle, the data show unambiguously that for concentrations in excess of 1 molal the angle of tilt, θ, between the Ni–O axis and the plane of the water molecule is substantial (figure 2 and table 2). As the concentration of NiCl$_2$ is

TABLE 1. SCATTERING LENGTH AND SAMPLE PARAMETERS (CATION HYDRATION)

electrolyte solution	isotopes	abundance (%)	scattering lengths 10^{-12} cm	c	molality	A/mb†	B/mb	C/mb	D/mb
NiCl$_2$.D$_2$O	^{N}Ni	—	1.03	0.0270	4.41	17.4	40.0	5.05	0.32
	^{62}Ni	94.9	−0.79	0.0192	3.05	12.6	29.0	2.52	0.15
				0.0093	1.46	06.4	14.6	0.61	0.04
				0.0056	0.85	03.85	08.85	0.22	0.013
				0.0028	0.42	01.94	04.46	0.054	0.00338
				0.00057	0.086	00.40	00.92	0.0023	0.00015
CaCl$_2$.D$_2$O	^{N}Ca	—	0.466	0.0275	4.49	03.0	06.8	0.09	0.02
	^{44}Ca	95.4	0.18						
LiCl.D$_2$O	^{6}Li	82.2	0.180	0.0277	3.57	03.47	07.92	0.85	−0.011
	^{7}Li	99.9	−0.233						

† 1 millibarn (mb) = 10^{-31} m^2.

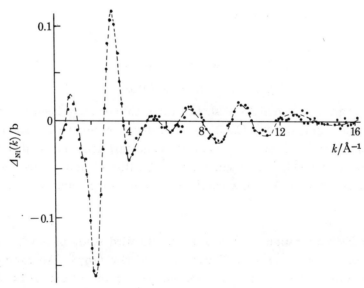

FIGURE 1. $\Delta_{Ni}(k)$ for a 4.41 molal solution of NiCl$_2$ in D$_2$O. The full circles represent experimental points and the smooth curve is the one used to calculate $G_{Ni}(r)$.

reduced, the Ni–O distance remains fixed, and although the errors in the measured value of the Ni–D distance become appreciable, the results strongly suggest that this distance increases. This provides clear structural evidence for the distortion of the hydration spheres as the packing fraction of the hydrated ions is increased; these observations, when combined with energy calculations, will allow a realistic estimate to be made of the repulsive part of the interionic potential. There are, so far as we are aware, no theoretical studies of the Ni–H$_2$O system from any of the standpoints discussed in §1. A neutron diffraction study of the crystal hydrate yields values for Ni–O, Ni–H and θ of 2.05 Å, 2.59 Å and 47°, respectively (Kleinberg 1969). These results emphasize the close similarity between concentrated solutions and their crystal hydrates (Friedman & Lewis 1976). The new feature of the solution data, i.e. the concentration

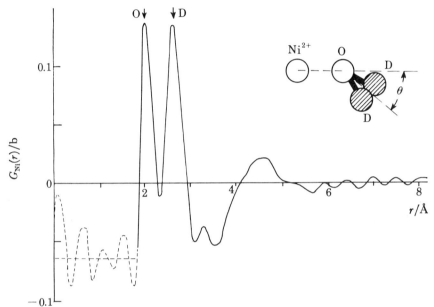

FIGURE 2. $G_{Ni}(r)$ for a 4.41 molal solution of NiCl$_2$ in D$_2$O.

TABLE 2. HYDRATION OF Ni^{2+}

solute	molality	ion–oxygen distance/Å	ion–deuterium distance/Å	θ/deg†	hydration number
NiCl$_2$	0.086	2.07 ± 0.03	2.80 ± 0.03	0 ± 20	6.8 ± 0.8
	0.46	2.10 ± 0.02	2.80 ± 0.02	17 ± 10	6.8 ± 0.8
	0.85	2.09 ± 0.02	2.76 ± 0.02	27 ± 10	6.6 ± 0.5
	1.46	2.07 ± 0.02	2.67 ± 0.02	42 ± 8	5.8 ± 0.3
	3.05	2.07 ± 0.02	2.67 ± 0.02	42 ± 8	5.8 ± 0.2
	4.41	2.07 ± 0.02	2.67 ± 0.02	42 ± 8	5.8 ± 0.2

† θ is the angle between the plane of the water molecule and the Ni–O axis.

TABLE 3. HYDRATION OF Ca^{2+}

solute	molality	ion–oxygen distance/Å	ion–deuterium distance/Å	θ/deg†	hydration number
CaCl$_2$	4.49	2.40 ± 0.03	2.93 ± 0.05	51 ± 15	5.5 ± 0.2

† θ is the angle between the plane of the water molecule and the Ca–O axis.

dependence of θ is, as yet, unexplained. A final comment to make on the Ni^{2+} data concerns the feature in $G_{Ni}(r)$ for $3.7 \leq r \leq 5.3$ Å. This is the first *structural* evidence for the existence of a *second* hydration shell; this shell becomes better defined as the concentration of $NiCl_2$ is reduced. The experimental resolution does not allow the ion–water configuration to be elucidated in detail but integration of $4\pi r^2 G(r)$ for $3.7 < r < 5.3$ Å yields a second hydration number of 15 ± 2.

TABLE 4. HYDRATION OF Li^+

solute	molality	ion–oxygen distance/Å	ion–deuterium distance/Å	θ/deg†	hydration number
LiCl	3.57	1.95 ± 0.02	2.55 ± 0.02	40 ± 10	5.5 ± 0.3

† θ is the angle between the plane of the water molecule and the Li–O axis.

3.1.2. Ca^{2+}

So far only one concentration of Ca^{2+} has been studied (Cummings *et al.* 1980) with the results shown in table 3. The ion–water conformation consistent with the data resembles that found for Ni and the measured Ca–O distance is exactly that given by molecular orbital calculations on an isolated ion–water pair (Kollman & Kuntz 1972). However, the molecular orbital calculations suggest that θ is zero for such a pair, in disagreement with these data. In the condensed phase, it is often suggested that a more favourable configuration may be one in which an oxygen lone pair points towards the Ca^{2+} ion and this model leads to a θ of *ca.* 60°, in reasonable agreement with the experimental value. It is clearly necessary to make measurements on more dilute solutions where this effect can be separated from that due to the distortion of the hydration spheres bought about by close packing; such an experiment has recently been completed (N. A. Hewish & J. E. Enderby, unpublished). The data, though not yet fully analysed, show that θ is *ca.* 40° at 1 molal. Thus the lone-pair argument, which should become increasingly valid at low c, is not supported by experiment so far as divalent cations are concerned. We therefore turn to the third cation studied, Li^+.

3.1.3. Li^+

Experiments on lithium salts (Newsome *et al.* 1980a) are complicated by the high absorption of ^6Li. A detailed study has shown that the ion–water distances are not dependent on the numerical procedures used to analyse data from highly absorbing samples like ^6LiCl in D_2O. There is, however, more uncertainty in the measured hydration number. The experiments on Li salts are of special interest because Li is a small, simple and monovalent ion. The results, summarized in table 4, show that θ is appreciably smaller than that expected from the lone pair configuration even at the high concentration used in the experiment. Once again, the importance of varying the concentration is evident but as it seems highly unlikely that an increase in θ would arise as c is *reduced*, we believe that the emphasis in much of the published work on the role of the lone pair electrons in deciding ion–water configurations in the condensed phase is in urgent need of revision. Our data at low concentrations for Ni^{2+} and Ca^{2+} support this conclusion. The hydration number, though subject to a large uncertainty at present, clearly favours *octahedral* rather than the *tetrahedral* coordination, a matter of long standing dispute in solution chemistry.

3.2. The hydration of anions

The method has been applied to Cl⁻ and in a wide range of solutions (table 5). Typical results for $\Delta_{Cl}(k)$ and $G_{Cl}(r)$ are shown in figures 3 and 4, where the Cl⁻–D$_2$O conformation consistent with these data is shown. A summary of all the results is presented in table 6. When we compare the nature of the Cl⁻ hydration among the various solutions, several conclusions emerge. First, the general form of $G_{Cl}(r)$ is remarkably similar in all cases. This apparent lack of sensitivity of anionic hydration to the nature of the counter-ion (with the exception of Ni²⁺)

TABLE 5. SCATTERING LENGTH AND SAMPLE PARAMETERS (ANION HYDRATION)

electrolyte solution	isotope	abundance (%)	scattering lengths 10⁻¹² cm	c	molality	A/mb	B/mb	C/mb	D/mb
LiCl	³⁵Cl	99.35	1.17	0.0588	9.95	16.6	38.3	−1.34	4.38
	³⁷Cl	90.4	0.35						
	³⁵Cl	99.35	1.17	0.0227	3.57	06.95	16.0	−0.2	0.6
	³⁷Cl	90.41	0.35						
NaCl	³⁵Cl	99.35	1.17	0.0331	5.32	09.9	22.7	0.65	1.37
	³⁷Cl	90.4	0.35						
RbCl	³⁵Cl	99.35	1.17	0.0192	2.99	05.8	13.3	0.52	0.46
	³⁷Cl	90.4	0.35						
CaCl$_2$	³⁵Cl	99.35	1.17	0.0275	4.49	16.0	36.8	1.16	3.77
	³⁷Cl	90.4	0.35						
NiCl$_2$	³⁵Cl	99.35	1.17	0.0270	4.35	15.7	36.3	2.40	3.60
	³⁷Cl	90.4	0.35						

TABLE 6. HYDRATION OF ANIONS OBTAINED BY NEUTRON DIFFRACTION

solute	molality	Cl–D(1)/Å	Cl–O/Å	Cl–D(2)/Å (range)	ψ/deg (range)	coordination number
LiCl	3.57	2.25±0.02	3.34±0.05	—	0	5.9±0.2
LiCl	9.95	2.22±0.02	3.29±0.04	3.50–3.68	0	5.3±0.2
NaCl	5.32	2.26±0.03	3.20±0.05	—	0–20	5.5±0.4
RbCl	4.36	2.26±0.03	3.20±0.05	—	0–20	5.8±0.3
CaCl$_2$	4.49	2.25±0.02	3.25±0.04	3.55–3.65	0–7	5.8±0.2
NiCl$_2$	4.35	2.29±0.02	3.20±0.04	3.40–3.50	22–32	5.7±0.2

emphasizes the importance of *local* effects. The coefficients of the partial radial distribution functions are more favourable in some cases than others, and the enhanced resolution thereby obtained allow us to define closely the Cl⁻–D$_2$O geometry. The data clearly favour the linear configuration tentatively proposed by Soper *et al.* (1977), but with a tilt angle generally less than 7°. The Cl–O distance is in good agreement with the value predicted by quantum mechanical cluster calculations (Schuster *et al.* 1975); it differs from that found in the early molecular dynamics simulation studies of Heinzinger (1976), which employed the so-called ST2 potential (Stillinger & Rahman 1974). It will be important to consider what refinements to this potential are needed if the simulation studies are to reflect more accurately the new data, and progress along these lines has been reported (Pàlinkàs *et al.* 1977). The exceptional case, Ni²⁺, clearly deserves further study; the fact that Ni²⁺ is a transition metal ion may have special significance, and indeed a dilution experiment recently performed (G. W. Neilson, unpublished) indicates that the tilt angle disappears at low c. Finally, it should be noted for the case of $CaCl_2$ that 5.8 ± 0.2 water molecules surround each Cl^- ion and 5.5 ± 0.2 water molecules surround each Ca^{2+} ion. Thus about 17 water molecules are instantaneously coordinated to each molecule of $CaCl_2$, which implies, for a 4.49 molal solution (ca. 11 water molecules per $CaCl_2$ molecule), substantial sharing of the water molecules between the cations and anions. The structural implications of this, particularly for the Ca–Cl radial distribution function, will form the basis of a future study. Water sharing is clearly significant for $NiCl_2$ solutions also, the only other case where both ions have been studied in the same solution.

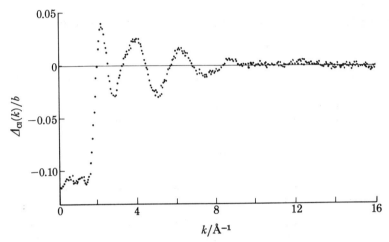

FIGURE 3. $\Delta_{Cl}(k)$ for a 9.95 molal solution of LiCl in D_2O. (J. R. Newsome, personal communication.)

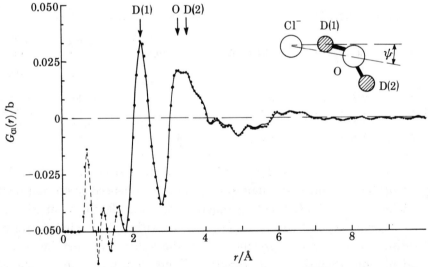

FIGURE 4. $G_{Cl}(r)$ for a 9.95 molal solution of LiCl in D_2O. (J. R. Newsome, personal communication.)

3.3. Ion–ion correlations

The first application of the second order difference method was made by Howe et al. (1974) to a 4.41 molal† solution of $NiCl_2$ in D_2O. The isotopic state of the Ni was changed so that data only for one of three partial structure factors, $S_{NiNi}(k)$, were obtained. $S_{NiNi}(k)$ was

† Not 5.51 molal as incorrectly stated in the original article.

characterized by (a) a well defined peak at a k of ca. 1 Å^{-1} and (b) considerable distortion beyond k of ca. 2 Å^{-1}, which was ascribed by the authors to the incomplete cancellation of the Plazek corrections. Neilson & Enderby (1980) repeated and extended the measurements to include Cl substitutions and eliminated the Plazek distortions by careful sample preparations in which the light water content of the samples was properly balanced. The results are shown in figures 5 and 6; the existence of the peak at 1 Å^{-1} is confirmed for the Ni–Ni case and is

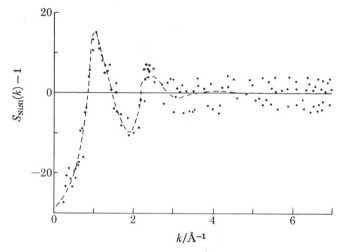

FIGURE 5. The partial structure factor, $S_{NiNi}(k)$ for a 4.41 molal solution of NiCl$_2$ in D$_2$O.

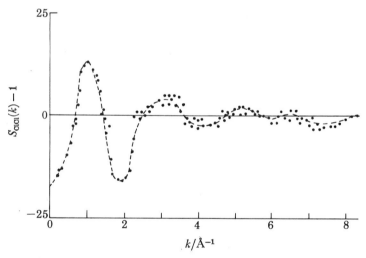

FIGURE 6. The partial structure factor, $S_{ClCl}(k)$ for a 4.41 molal solution of NiCl$_2$ in D$_2$O.

clearly demonstrated for the Cl–Cl case. However, these experiments are at the very limit of the existing technique and the results are not yet of sufficient accuracy to derive a fully self-consistent $g_{NiNi}(r)$ or $g_{ClCl}(r)$. We shall therefore restrict our discussion to k-space, although it is hoped in future to improve the basic technique and thereby allow data of higher quality to be obtained.

In their original article, Howe et al. (1974) argued that the data implied that the ions were arranged in a more ordered way than would be expected from primitive models. Experimental support for this view came from a variety of sources (March & Tosi 1974; Fontana 1976; Cubiotti et al. 1977), and was further strengthened by the observation that the position k_0 of the prepeak in $F(k)$ the *total* diffraction pattern, characteristic of $NiCl_2$ solutions, scaled to the one-third power of the molarity (Neilson et al. 1975). This approach was challenged by Quirke & Soper (1977) who showed that by representing the Ni^{2+} and Cl^- ions (appropriately 'dressed' by six water molecules) as hard spheres, substantial values of $S_{NiNi}(k)$ at the first maximum could be obtained. It is now clear, however, that hard-sphere structure factors are

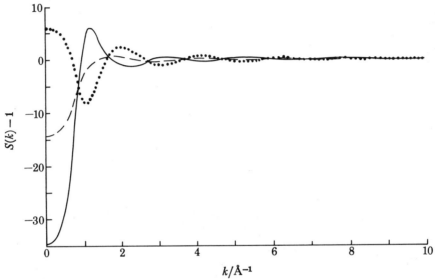

FIGURE 7. Partial structure factors derived for a primitive model of a molal solution of $NiCl_2$. The diameter was chosen as 3.0 Å and the dielectric constant taken as 60. Full curve, $S_{NiNi}(k)$; broken curve, $S_{ClCl}(k)$; dotted curve, $S_{NiCl}(k)$ (M. Telo da Gama & R. E. Evans, private communication).

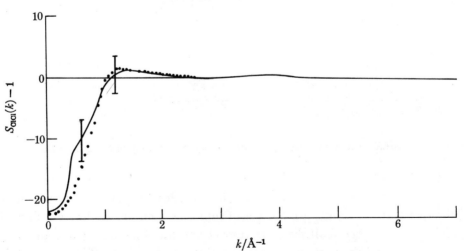

FIGURE 8. The partial structure factor $S_{ClCl}(k)$ for a 5.32 molal solution of NaCl in D_2O. Full curve, experiment; dotted curve, theory (mean spherical approximation).

vastly changed once Coulomb interactions are included (see especially Enderby & Neilson 1980). An analytic and instructive approach to the evaluation of ion–ion structure factors within the primitive model is the so-called mean spherical approximation (Waisman & Lebowitz 1972) whose properties have been recently explored in this laboratory by Telo da Gama & Evans (private communication).

The k-space results derived from the work of Telo da Gama & Evans for a 4 molal solution of $NiCl_2$ in D_2O with an effective ionic diameter chosen to be 3.0 Å are shown in figure 7. Clearly this approach fails to predict the marked peaks found in our experiments for the *like* distribution functions. The newer and more reliable data therefore support the original conclusions of Howe *et al.* (1974) and March & Tosi (1974), that primitive models either with or without Coulomb interactions are inconsistent with the experimental facts so far as $NiCl_2$ is concerned. A series of investigations has been recently completed for Cl^- ions in a range of 1–1 systems is in progress (Newsome *et al.* 1980b). In the only case so far analysed in detail (a 5.32 molal solution of NaCl), the experimental data agree in their general form with those predicted by Telo da Gama & Evans (see figure 8). It is our view that the molecular nature of the water is a crucial factor in sustaining long-range effects in certain aqueous solutions and an extension of the approach developed by Outhwaite (1976) and others is a necessary step towards a more reliastic theory. In other cases, the primitive model appears to contain all the essential physics. The distinction arises, we believe, from *intermediate*-range chemical effects, i.e. those that cannot be incorporated in the primitive model by adjusting the ion size. In order to identify these ions, for which intermediate-range chemical effects are important, a more systematic investigation is needed and we turn finally to consider what experiments could form part of a future research programme.

TABLE 7

valence	ion	possible isotopes		scattering lengths known	feasibility	
					hydration	ion–ion correlations
1	Li^+	^6Li	^7Li	yes	2	3
1	$(ND_4)^+$	^{14}N	^{15}N	yes	2	4
1	Ag^+	^{107}Ag	^{109}Ag	yes	2	4
1, 2	Cu^+, Cu^{2+}	^{63}Cu	^{65}Cu	yes	1	3
2	Mg^{2+}	^{24}Mg	^{25}Mg	yes	2	4
2	Ca^{2+}	^{40}Ca	^{44}Ca	yes	1	2
2	Ni^{2+}	^{58}Ni	^{60}Ni ^{62}Ni	yes	1	2
2	Ba^{2+}	^{138}Ba	^{137}Ba	no	?	?
2	Sr^{2+}	^{86}Sr	^{88}Sr	no	?	?
2	Zn^{2+}	^{64}Zn	^{68}Zn	yes	4	5
2	Sn^{2+}	^{102}Sn	^{122}Sn	yes	4	5
2	Hg^{2+}	^{200}Hg	^{202}Hg	no	?	?
2, 3	Fe^{2+}, Fe^{3+}	^{56}Fe	^{57}Fe	yes	1	2
2, 3	Cr^{2+}, Cr^{3+}	^{52}Cr	^{53}Cr	yes	1	2
-1	Cl^-	^{35}Cl	^{37}Cl	yes	1	2
-1	I^-	^{127}I	^{129}I†	no	?	?
-1	$(NO_2)^-, (NO_3)^-$	^{14}N	^{15}N	yes	1	3
-1	$(CN)^-, (SCN)^-$	^{14}N	^{15}N	yes	1	2
-1	$(ClO_3)^-, (ClO_4)^-$	^{35}Cl	^{37}Cl	yes	1	2
-2	$(SO_3)^-, (SO_4)^-$	^{32}S	^{33}S	yes	3	4
-2	$(CO_3)^-$	^{12}C	^{13}C	yes	4	5

† Slightly radioactive.

4. FUTURE PROSPECTS

The principles behind the technique are now well established and we can look to the future. A careful study of nuclear isotopes, their level structure and their availability shows that, contrary to what is frequently supposed, a wide but exemplary series of experiments is possible. There are *at least* 14 cations and 11 anions for which appropriate isotopic substitutions can in principle be made, and these are shown in table 7. The feasibility of any given experiment depends on a combination of scattering lengths, salt concentrations and linear absorption coefficients. We are now able to state explicitly which of all the possible experiments are feasible. For simplicity, the results are expressed in table 7 on a scale of 1 to 5 (1, feasibility demonstrated, experiment well understood; 5, not possible with present technology). For practical purposes, only ions with a feasibility number in the range 1–3 need be considered at present, although an improvement in effective neutron counting rates by a factor of 10 would allow category 4 experiments to be considered. It is apparent that an extensive programme of research designed to answer many of the longstanding but specific questions in this subject is implied by Table 7. One might, for example, compare the conformation of water with respect to Fe^{2+} and Fe^{3+} ions; the nature of the hydration around *nitrate ions* is another topic that would repay investigation. A careful study of Mg^{2+} with a view to its isomorphic replacement by Ni^{2+} is yet another line of attack. So far as ion–ion correlations are concerned, important candidates for study are those for which isotopes of both cations and anions can be changed so that all three correlation functions, including the important cross-term, can be measured. Some examples are, for 1–1 systems, $LiCl$ and ND_4NO_3; for 2–1 systems, $CaCl_2$ and $MgCl_2$; and for 3–1 systems, $FeCl_3$ and $CrCl_3$. With the development of new and powerful sources of thermal neutrons like, for example, the Rutherford Laboratory's Spallation Source, the future prospects for a major programme of research into the fundamental structure of solutions look extremely favourable.

The experimental work described in this paper has been generously supported by the Science Research Council. I wish to thank the Council, past and present members of the Bristol/Leicester solutions group and the staff at the I.L.L. for making possible the advances that I have described.

REFERENCES (Enderby)

Cubiotti, G., Maisano, G., Migliardo, P. & Wanderlingh, F. 1977 *J. Phys.* C **10**, 4689.
Cummings, S., Enderby, J. E. & Howe, R. A. 1980 *J. Phys.* C. **13**, 1.
Enderby, J. E. & Neilson, G. W. 1980 *Adv. Phys.* **29**, 323.
Fontana, M. P. 1976 *Solid State Commun.* **19**, 765.
Friedman, H. L. & Lewis, L. 1976 *J. Soln Chem.* **5**, 445.
Heinzinger, K. 1976 *Z. Naturf.* A **31**, 1073.
Howe, R. A., Howells, W. S. & Enderby, J. E. 1974 *J. Phys.* C **7**, L111.
Kleinberg, R. 1969 *J. chem. Phys.* **50**, 4690.
Kollman, P. & Kuntz, I. D. 1972 *J. Am. chem. Soc.* **94**, 9236.
March, N. H. & Tosi, M. P. 1974 *Phys. Lett.* A **50**, 224.
Neilson, G. W. & Enderby, J. E. 1978 *J. Phys.* C **11**, L625.
Neilson, G. W. & Enderby, J. E. 1980 (In preparation.)
Neilson, G. W., Howe, R. A. & Enderby, J. E. 1975 *Chem. Phys. Lett.* **33**, 284.
Newsome, J. R., Neilson, G. W. & Enderby, J. E. 1980a *J. Phys.* C. (Submitted.)
Newsome, J. R., Soper, A. K., Neilson, G. W. & Enderby, J. E. 1980b (In preparation.)
Outhwaite, C. W. 1976 *Molec. Phys.* **31**, 1345.

Pàlinkàs, G., Riede, W. O. & Heinzinger, K. 1977 *Z. Naturf.* A **32**, 1137.
Quirke, N. & Soper, A. K. 1977 *J. Phys.* C **10**, 1802.
Schuster, P., Jakubetz, W. & Marius, W. 1975 *Top. curr. Chem.* **60**, 1.
Soper, A. K., Neilson, G. W., Enderby, J. E. & Howe, R. A. 1977 *J. Phys.*, C **10**, 1793.
Stillinger, F. H. & Rahman, A. 1974 *J. chem. Phys.* **60**, 1545.
Waisman, E. & Lebowitz, J. L. 1972 *J. chem. Phys.* **56**, 3093.

Discussion

E. W. J. MITCHELL (*Clarendon Laboratory, Oxford, U.K.*). Lock, Missolloras, Mitchell & Stewart have repeated the measurements of the partial structure factors of molten CsCl by using the isotope substitution method, with vanadium containers rather than the silica ones used by Derrien & Dupuy. The partial $g(r)$'s have been determined and are in good agreement with the molecular dynamics simulation of Dixon & Sangster (1977) as in their fig. 2. The main features (experimental values followed by theoretical values in parentheses) are:

g_{+-}	first peak	position	3.37 Å	(3.3)
		height	4.3	(4.1)
	first minimum	position	5.4	(5.0)
g_{--}	first peak	position	4.67	(4.6)
		height	1.63	(1.7)
	subsid. peak	position	6.76	(6.6)
		height	1.09	(1.1)

The full analysis is in course of publication. We do not observe the additional features found by Derrien & Dupuy.

Reference

Dixon, M. & Sangster, M. J. L. 1977 *J. Phys.* C **10**, 3015.

J. G. POWLES (*University of Kent at Canterbury, U.K.*). Professor Enderby has chosen to discuss in detail his important work on the structure of aqueous electrolytic solutions exploiting the isotope substitution method. It is perhaps worth remarking, as he is well aware, that the neutron scattering technique, and the isotope substitution method in particular, has made possible important advances in our understanding of the whole field of liquids and amorphous systems. It has been applied also with success, in particular, to molten salts (Page & Mika 1971), to molecular liquids (Bertagnolli *et al.* 1978) and to compressed gases (Soper & Egelstaff 1980), and there is no doubt that in due course we shall have very detailed and accurate information as to the structure of a great variety of fluid systems over a wide range of state conditions. This presents a serious challenge to the theory of fluids, and indeed great progress has been made in recent years in our understanding of the structure and dynamics of fluids.

It should be noted that the isotope substitution method is very demanding of instrument time wherever more than two or three different nuclei are present, even for one state point, and this further emphasizes the necessity of improved facilities especially as regards neutron flux. Nevertheless it should not be overlooked that even one structure factor for a given liquid can be extremely helpful in defining the structure of a fluid, and of course it is quite economical. In many important cases there may be only one partial structure factor anyway, e.g. fluids of homonuclear diatomic molecules, and in many cases some of the partial structure factors make only a small contribution to the total. Thus in liquid VCl$_4$ only S_{Cl-Cl} is significant (Gibson & Dore 1979). The important contribution of an X-ray structure factor should also not be overlooked.

Although, generally speaking, the coherent scattering length of most important isotopes are known there are still some gaps and some uncertainties (the b value for D was substantially changed recently!). The measurement of absolute scattering cross section is still tedious and difficult but is an important check on the analysis of data and shows the necessity of even more accurate measurement of scattering lengths.

The incoherent and the absorption cross sections are much less well known and this is important since the effect of these has to be allowed for very accurately before the coherent scattering law can be extracted, to sufficient accuracy, from the experimental data.

There is also the important problem of the recoil and detector corrections for light, and even not so light, nuclei (Powles 1979), which was neatly circumvented by Professor Enderby, but which in general presently tends to limit quite severely the accuracy of extracted coherent structure factors for many important liquids, e.g. water. This problem should also be eased by the availability of adequate fluxes of epithermal neutrons from new facilities such as the S.N.S.

It may be pointed out also that the slow neutron scattering data on liquids, and particularly on molecular liquids, have provided important information on the interatomic or intermolecular potentials (Cheung & Powles 1976), which are extremely difficult to determine by any method. This information is of course basic to our understanding of the condensed state, including crystals. The second virial coefficient method has been recently extended to neutron scattering (Powles *et al.* 1979), but here again a thorough investigation will demand the availability of much more abundant supplies of neutrons than are currently available. This seems likely to resurrect the field of neutron scattering by dilute gases, which has been rather neglected in recent years.

Finally, it may be noted that the study of the dynamics of disordered systems is developing rapidly and we may expect a major effort in this field in the next few years now that we are approaching a thorough understanding of the structural properties, which is an essential prerequisite.

References

Bertagnolli, H., Leicht, D. O., Zeidler, M. D. & Chieux, P. 1978 *Molec. Phys.* **36**, 1769, and earlier papers from this group.
Cheung, P. S. Y. & Powles, J. G. 1976 *Molec. Phys.* **32**, 1383, and earlier papers.
Gibson, I. P. & Dore, J. C. 1979 *Molec. Phys.* **37**, 1281.
Page, D. I. & Mika, K. 1971 *J. Phys.* C **4**, 3034.
Powles, J. G. 1979 *Molec. Phys.* **37**, 623, and earlier papers.
Powles, J. G., Dore, J. C. & Osae, E. K. 1980 *Molec. Phys.* **40**, 193.
Soper, A. K. & Egelstaff, P. A. 1980 *Molec. Phys.* **39**, 1201.

Phil. Trans. R. Soc. Lond. B **290**, 567–582 (1980)

Molecular crystals and liquid crystals: new results for *t*-butyl chloride

By J. C. Frost†‡, A. J. Leadbetter† and R. M. Richardson§
† *Chemistry Department, University of Exeter, Exeter EX4 4QD, U.K.*
§ *Rutherford Laboratory, Chilton, Didcot, Oxon. OX11 0QX, U.K.*

The main theme of this paper is the use of incoherent quasielastic scattering to study molecular reorientational motions in molecular and liquid crystals at temperatures where a classical description is appropriate. The relation to quantum mechanical tunnelling between orientational potential wells of hydrogenous species (e.g. CH_3, NH_4^+) is mentioned briefly.

The use of this technique is illustrated with a description of the determination of the nature and dynamics of the molecular reorientations in the solid phases of *t*-butyl chloride. This shows how the measurement of the quasielastic spectrum as a function of scattering vector permits an identification of the type of reorientation and its characteristic time scale, provided this is fast enough to be observed in a neutron experiment ($\tau \gtrsim 10^{-9}$ s).

Finally a brief résumé is given of the results that have emerged from such studies about the various molecular motions that characterize and distinguish different liquid crystal phases.

1. Introduction

Very many substances have one or more phases in which the molecules are orientationally disordered. This disorder must in general be dynamic and is important in determining the properties of the phase. Furthermore, the understanding of the transitions leading to the formation of these phases is a challenging theoretical problem. A full understanding of these systems must necessarily involve some cooperative effects between molecules, and these may be quite dominant. Nevertheless a detailed understanding of the single molecule properties may yield considerable information about the orientational potential and the nature of the transitions. Provided that the molecule contains a sufficient fraction of protons, then incoherent neutron scattering experiments provide a very powerful means of achieving this. The proton has a very high and incoherent scattering cross section (*ca.* 80 b¶ compared with a few barns for most other nuclei; e.g. *ca.* 5 b for C) so that, given enough protons, the incoherent scattering from these dominates the observed scattering and renders its interpretation relatively straightforward.

Another area in which such methods are very useful is in the study of intramolecular reorientations. In both cases there are two distinct experimental aspects: the inelastic spectrum gives information about the torsional or librational levels and the quasielastic spectrum about the stochastic jumping between different potential wells. The barrier heights between pocket states may be so weak compared with the thermal energy that the motion goes over to rotational diffusion with a more or less random distribution of orientation, but this is uncommon (Leadbetter & Lechner 1979; Hüller & Press 1978; Stiller 1978).

‡ Present address: Chemistry Department, University of California, Berkeley, California 94720, U.S.A.
¶ 1 barn (b) = 10^{-28} m².

At sufficiently low temperatures and for small enough moments of inertia, the rotational motions must be described in terms of quantum mechanical tunnelling states. In practice this tunnelling occurs only for H and D atoms so that only groups such as CH_4, NH_4^+ and $-CH_3$ are involved. The study of such tunnelling states is important for two reasons. First, they provide additional parameters for the detailed determination of the rotational potential, and secondly,

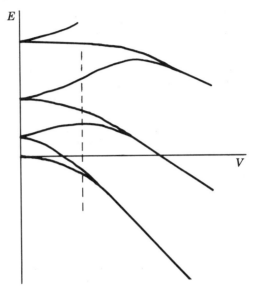

FIGURE 1. Energy level correlation diagram showing the relation between free rotation and vibrational levels for a system in an orientational potential of height V. The dashed line shows schematically a potential below which the tunnel splittings might be observable in a neutron experiment.

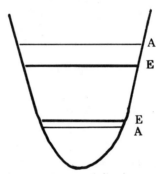

FIGURE 2. Energy diagram for threefold rotational system (e.g. $-CH_3$), showing the first two torsional levels with their tunnel splittings.

they are necessary for an *explanation* of the reorientational processes compared with their mere description (Hüller & Press 1978, Clough & Heidemann 1979; Clough *et al.* 1978). Although these studies are not the subject of this paper, a brief discussion will be given because of their great importance to the understanding of rotational motions and because neutrons have recently come to play a major role in this work. The nature of the tunnelling states may be seen with reference to the schematic energy correlation diagram of figure 1. In the extreme of low orientational potential, the states are those for quantum mechanical free rotation while at the extreme of high potential they are librational (harmonic) oscillator states. The splitting of the librational states is the tunnel splitting and in many interesting cases it falls within the range accessible to neutron studies ($\Delta E \gtrsim 1$ μeV typically for $V \lesssim 400$ cm^{-1} ≈ 50 meV. The situation is shown in figure 2 for the $-CH_3$ case where the two lowest libration states are shown. Typical spectra for the ground state splitting as a function of T for this case are shown in figure 3 (after Batley et al. 1977), which shows how the tunnelling picture goes over at still relatively low temperature to one where a classical stochastic description of the reorientation is adequate.

A number of very elegant experiments have recently been performed on the study of the temperature and pressure dependence of tunnelling splittings and line widths for $-CH_3$ rotations, including measurements of the first excited torsional state (Clough & Heideman 1979; Clough et al. 1978, 1979) as well as a more complicated three-dimensionally rotating system (Prager et al. 1977; Press & Prager 1977; Prager & Alefeld 1976).

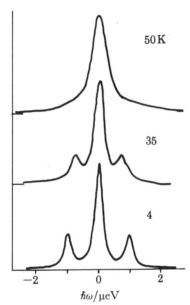

FIGURE 3. Typical neutron scattering tunnelling spectra (after Batley et al. 1977) for a $-CH_3$ group as a function of temperature, showing the change from tunnelling transitions at low T to classical quasielastic broadening at higher T.

2. INCOHERENT NEUTRON SCATTERING FROM MOLECULES UNDERGOING BOUND MOTIONS

A detailed review of this topic has recently been published by Leadbetter & Lechner (1979), so here it is necessary only to summarize a few important results.

First it is assumed that the molecules contain sufficient protons so that the total scattering from them is dominated by the incoherent cross section of the protons (except at Bragg reflections). The incoherent scattering law is then the time Fourier transform,

$$S_s(\boldsymbol{Q}, \omega) = \tfrac{1}{2}\pi \int_{-\infty}^{\infty} \{\exp(-i\omega t)\} I_s(\boldsymbol{Q}, t) \, dt,$$

of the intermediate scattering function

$$I_s(\boldsymbol{Q}, t) = \exp\{i\boldsymbol{Q} \cdot \langle [\boldsymbol{r}(t) - \boldsymbol{r}_0(0)] \rangle\},$$

which may alternatively be written in terms of the conditional probability function $G_s(\boldsymbol{r}, \boldsymbol{r}_0, t)$ and the probability distribution $g(\boldsymbol{r}_0)$ of initial positions $\boldsymbol{r}_0$,

$$I_s(\boldsymbol{Q}, t) = \iint \exp i\boldsymbol{Q} \cdot [\boldsymbol{r} - \boldsymbol{r}_0] \, G_s(\boldsymbol{r}, \boldsymbol{r}_0, t) \, g(\boldsymbol{r}_0) \, \mathrm{d}\boldsymbol{r} \, \mathrm{d}\boldsymbol{r}_0.$$

When the molecule contains dynamically non-equivalent protons, $S_s(\boldsymbol{Q}, \omega)$ must be averaged over them to give the scattering law for the molecule.

For bound motions such that the proton (molecule) is confined to a restricted volume of space, the long time limit of $G_s(\boldsymbol{r}, \boldsymbol{r}_0, t) = G_s(\boldsymbol{r}, \infty)$ may be separated out to give

$$I_s(\boldsymbol{Q}, t) = I_s(\boldsymbol{Q}, \infty) + I_s'(\boldsymbol{Q}, t),$$

and since $G_s(\boldsymbol{r}, \infty) = g(\boldsymbol{r}_0)$,

$$I_s(\boldsymbol{Q}, \infty) = |\int \{\exp(i\boldsymbol{Q} \cdot \boldsymbol{r})\} G_s(\boldsymbol{r}, \infty) \, \mathrm{d}\boldsymbol{r}|^2 \quad (1)$$
$$= |\langle \exp(i\boldsymbol{Q} \cdot \boldsymbol{r}) \rangle|^2$$
$$= A_0(\boldsymbol{Q}),$$

and the scattering law becomes

$$S_s(\boldsymbol{Q}, \omega) = A_0(\boldsymbol{Q}) \, \delta(\omega) + S_s'(\boldsymbol{Q}, \omega), \quad (2)$$

where $A_0(\boldsymbol{Q})$ is called the elastic incoherent structure factor (e.i.s.f.).

For a vibrational motion of frequency ω_0, (2) becomes

$$S_s(\boldsymbol{Q}, \omega) = \exp \langle (\boldsymbol{Q} \cdot \boldsymbol{u})^2 \rangle \, \delta(\omega) + S_s^{\mathrm{inel}}(\boldsymbol{Q}, \omega), \quad (3)$$

where $A_0(\boldsymbol{Q})$ is now the well known Debye–Waller temperature factor and $S_s^{\mathrm{inel}}(\boldsymbol{Q}, \omega)$ comprises peaks at $\pm \hbar \omega_0$.

For bound stochastic motion, for example rotational motion about the centre of mass

$$S_s(\boldsymbol{Q}, \omega) = A_0(\boldsymbol{Q}) \, \delta(\omega) + S_s^{\mathrm{qe}}(\boldsymbol{Q}, \omega), \quad (4)$$

where $A_0(\boldsymbol{Q})$ now depends on the geometrical nature of the rotations and examples are shown by the data in figures 6 and 7. $S_s^{\mathrm{qe}}(\boldsymbol{Q}, \omega)$ is a quasielastic scattering component that depends upon both the geometry and timescale of the rotations. Usually $I_s'^R(\boldsymbol{Q}, t)$ will be of the general form

$$I_s'^R(\boldsymbol{Q}, t) = \sum_{l=1} A(\boldsymbol{Q}) \exp(-t/\tau_l),$$

so that

$$S_s^{\mathrm{qe}}(\boldsymbol{Q}, \omega) = \sum_{l=1} A_l(\boldsymbol{Q}) \, \mathscr{L}(1/\tau_l), \quad (5)$$

where

$$\mathscr{L}(1/\tau_l) = \pi^{-1} \tau_l^{-1} / (\tau_l^{-2} + \omega^2).$$

Here we shall be concerned mainly with such motions so that the observed scattering law will be of the form of (4) and (5). Typical examples are shown by the data of figures 4 and 5.

Detailed discussion of various examples may be found in Leadbetter & Lechner (1979) but we give here for illustration the results for three simple cases, because they are relevant to the case of t-butyl chloride (and the liquid crystals) discussed below.

Threefold jump reorientation

For a single crystal specimen,

$$S_s(\boldsymbol{Q}, \omega) = [\tfrac{1}{3} + \tfrac{2}{9}(\cos \boldsymbol{Q} \cdot \boldsymbol{r}_{12} + \cos \boldsymbol{Q} \cdot \boldsymbol{r}_{23} + \cos \boldsymbol{Q} \cdot \boldsymbol{r}_{13})] \, \delta(\omega) + [\tfrac{2}{3} - \tfrac{2}{9}(\cos \boldsymbol{Q} \cdot \boldsymbol{r}_{12}$$
$$+ \cos \boldsymbol{Q} \cdot \boldsymbol{r}_{23} + \cos \boldsymbol{Q} \cdot \boldsymbol{r}_{13})] \, \mathscr{L}(\tfrac{2}{3}\tau), \quad (6)$$

where τ is the residence time (τ^{-1} the jump probability) and 1, 2 and 3 label the sites. For a powder this becomes (for equidistant sites)

$$S_s(Q, \omega) = [\tfrac{1}{3} + \tfrac{2}{3}j_0(Qa\sqrt{3})]\,\delta(\omega) + (\tfrac{2}{3} - \tfrac{2}{3}j_0(Qa\sqrt{3}))\,\mathscr{L}(\tfrac{2}{3}\tau), \tag{7}$$

where a is the radius of the circle on which the sites are located and $j_0(x)$ is the zero order spherical Bessel function ($j_0(x) = x^{-1}\sin x$). Most quasielastic studies of molecular reorientations so far have been made on powder samples, but notable exceptions are studies on orientated liquid crystal samples described briefly below.

Rotation on a circle

The above result (equation (7)) has been extended to the general case of m equidistant sites on a circle of radius a (Barnes 1973) for a powder sample, when

$$S_s(Q, \omega) = A_0(Q)\,\delta(\omega) + \sum_{l=1}^{l=m} A_l(Q)\,\mathscr{L}(\tau_l^{-1}), \tag{8}$$

where
$$A_l(Q) = m^{-1} \sum_{j=1}^{m} j_0\{2Qa\sin(\pi j/m)\}\cos(2\pi lj/m) \tag{9}$$

and
$$\tau_l = \tfrac{1}{2}\tau/\{\sin^2(\pi l/m)\}. \tag{10}$$

For $m \to \infty$, these results are equivalent to continuous rotational diffusion on the circle with $D_r = \tau_1^{-1}$ (Dianoux et al. 1975).

Rotational diffusion on a sphere of radius a

$$S_s(Q, \omega) = j_0^2(Qa)\,\delta(\omega) + \sum_{l=1}^{\infty}(2l+1)\,j_l^2(Qa)\,[l(l+1)\,D_r]. \tag{11}$$

D_r is the rotational diffusion constant and j_0 a spherical Bessel function.

Translational diffusion

When continuous centre-of-mass diffusion occurs the scattering law becomes a Lorentzian of argument D_tQ^2 where D_t is the translational diffusion coefficient. Because low Q data probe the large distance components of motion ($A_0(Q) \to 1$ as $Q \to 0$), then measurements at low enough Q enable D_t to be determined even if the molecule is also rotating. The separation of the rotational motion from translation at high Q depends on the relative magnitudes of D_tQ^2 and τ_l^{-1}: the effect of translational motion being to broaden, by folding with $\mathscr{L}(D_tQ^2)$, the scattering laws discussed above. Examples of cases where such separation is possible will be discussed below for liquid crystals.

3. Tertiary butyl chloride, $(CH_3)_3CCl$

3.1. *Introduction*

This is an excellent example for illustrating the use of high resolution incoherent neutron quasielastic scattering (i.n.q.e.s.) techniques because, despite a great amount of work, particularly with other techniques, the nature of the molecular disorder in the three solid phases is still not clearly established. The phase behaviour is as follows:

$$\text{crystal III} \underset{}{\overset{183\text{ K}}{\rightleftarrows}} \text{crystal II (tetragonal)} \underset{}{\overset{219\text{ K}}{\rightleftarrows}} \text{crystal I (cubic)} \underset{}{\overset{247\text{ K}}{\rightleftarrows}} \text{liquid}.$$

Crystal III is an ordered phase whose detailed structure has not yet been published. Crystal II is tetragonal (Rudman & Post 1968), and crystal structure determination suggested a head-to-tail packing of molecules having collinear C–Cl and random orientation of methyl groups about this axis. Crystal I is f.c.c. with the molecules having non-random but undetermined orientations (Schwartz et al. 1951). These structural results are in accord with dielectric (Kushner et al. 1950) and n.q.r. (O'Reilly et al. 1973) measurements which indicate tumbling of the C–Cl axis in I but not in II and with calculations and for i.r. absorption data of Lassier & Brot (1968) suggesting favoured sites in I comprising a manifold of cubic symmetry. Some recent relatively low resolution i.n.q.e.s. results (Goyal et al. 1974) on phase II were interpreted as most consistent with threefold reorientations about the dipole axis, whereas later measurements (Larsson et al. 1978; Mansson & Larsson 1977) were interpreted as showing rotation about all axes in II as well as I, although that about the dipole axis was some 9 times faster than in the other directions, all molecular rotation being frozen in III. This result for II disagrees with deductions from other experiments (above).

A very detailed nuclear magnetic resonance study was carried out by O'Reilly et al. (1973) in which it was concluded, on the basis of spin lattice relaxation (T_1 and $T_{1\rho}$) measurements, that in solid III the rates of methyl reorientations about their symmetry axes obey an Arrhenius law: $\tau = \tau_0 \exp(E_a/kT)$ with $E_a = 15.2$ kJ mol^{-1} and $\tau_0 = 7.3 \times 10^{-15}$ s. Another motion on a time scale of 10^{-7} to 10^{-8} s, in crystal II and III, was identified as a reorientation of the whole molecule about the C–Cl axis. These conclusions are inconsistent with the results of detailed i.n.q.e.s. and neutron inelastic (i.n.i.s.) scattering experiments on the closely related compound, t-butyl cyanide (Frost 1979; Frost et al. 1980). For this substance the molecules reorientate about the dipolar axis even in the lowest temperature phase with $\tau \approx 10^{-10}$ s, and as the temperature increases this process rapidly becomes much faster than the methyl reorientation which, from the methyl torsion frequencies and direct i.n.q.e.s. observations, can be characterized by an Arrhenius law ($E_a = 16.3$ kJ mol^{-1}, $\tau_0 = 0.20 \times 10^{-12}$ s). As both compounds have similar methyl torsional frequencies (Durig et al. 1969; Ratcliffe & Waddington 1976) the methyl reorientation correlation times should be similar and hence much longer than those deduced from the n.m.r. experiments, which also give an unphysically high value for τ_0.

We have therefore made new measurements at 165, 202 and 236 K, i.e. one measurement in each solid phase, using the time-of-flight multi-chopper spectrometer (IN5) at the Institut Laue–Langevin, Grenoble. The sample was contained in a stainless steel can and had a thickness of ca. 0.3 mm. The incident wavelength of 7.8 Å† gave the experiment an elastic momentum transfer range of 0.35 to 1.47 Å^{-1} and an elastic energy resolution (f.w.h.m.) of ca. 31 μeV. The spectra obtained were transformed by using well known procedures to the scattering law $S_s(Q, \omega)$. The results for crystal III showed no quasielastic broadening indicating that any random molecular motions (intramolecular or intermolecular reorientations) are slower than 3×10^{-10} s. The results for crystals II and I are shown respectively in figures 4 and 5 at three different Q values and the experimental e.i.s.f. data are plotted in figures 6 and 7. These were obtained by using a least squares computer fit and equations of the form of (4) and (5). The determination of reliable e.i.s.f. values from IN5 is facilitated because of the triangular shape of the resolution function.

† $1 \text{ Å} = 10^{-10}$ m $= 10^{-1}$ nm.

3.2. Crystal II

The spectra for II clearly show an elastic peak superimposed on a broadened component with a full width at half maximum of 360 µeV (implying a correlation time of ca. 4×10^{-12} s). The slower motion observed by the n.m.r. experiment in this phase (and attributed to uniaxial molecular reorientation) is far too slow to give any detectable broadening, as is translational diffusion. Because of instrumental difficulties in measuring $T_{1\rho}$, O'Reilly et al. (1973) could not determine correlation times for the faster motion in this phase, but from their data in solid I and III this time must lie between ca. 10^{-10} and ca. 6×10^{-12} s. It seems certain therefore that the n.m.r. and i.n.q.e.s. experiments relate to the same fast motion but the experimental e.i.s.f. shows that this motion is definitely *not* a reorientation of the methyl groups as suggested by O'Reilly et al. Instead it is much more closely related to rotational diffusion of the whole

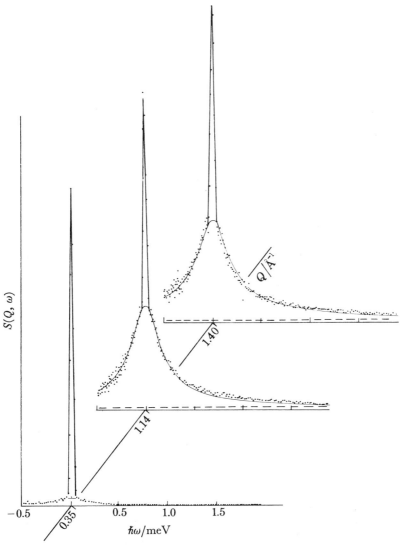

FIGURE 4. Typical neutron spectra for the crystal II phase of t-butyl chloride at 202 K. The full line represents a fit to the data of the uniaxial rotation model (equation (8); $m \gtrsim 6$), but with the coefficient of the elastic peak a fittable parameter. This gives the e.i.s.f. of figure 6. The dashed line shows the inelastic background.

molecule about the C–Cl axis. The e.i.s.f. for the model of simple uniaxial rotational diffusion ($m \gtrsim 6$ (equation (8)), with the use of known molecular geometry) is shown in figure 6.

The experimental e.i.s.f. is in fact lower than that for uniaxial rotational diffusion, even when multiple scattering is taken into account (figure 6) which was done with the use of a Monte Carlo method (Johnson 1974). This shows that the motion must be even less restricted

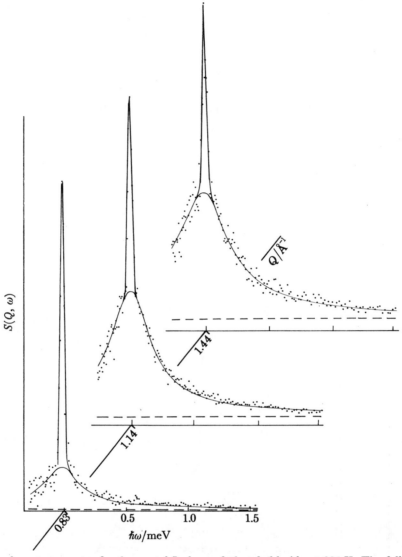

FIGURE 5. Typical neutron spectra for the crystal I phase of t-butyl chloride at 236 K. The full line shows the fit to the data of the model for random rotational diffusion on a sphere (equation (11)).

than just random rotation about the C–Cl axis. Reorientations of the methyl groups might cause this but in fact they are too slow to make a significant contribution to the quasielastic broadening and hence to have been seen in the present experiments (see below). Therefore there must be motions of the molecule about its other axes, although these must be much more localized than overall rotation. Another possibility is that as a molecule lacks the room to rotate freely about any axis, and the molecular and (average) site symmetries are different (Lassier & Brot 1968), then such a reorientation may be accompanied by a lateral relaxation of the neighbouring molecules. The e.i.s.f. may be fitted by a variety of similar models for the limited Q range of these experiments and it is not possible to distinguish between them without additional evidence. Two examples are shown in figure 4: (i) a small step uniaxial diffusion (equation (8) with $m \gtrsim 6$) plus a random lateral displacement of the protons with $\langle x^2 \rangle^{\frac{1}{2}} \approx 0.7$ Å; (ii) threefold uniaxial rotation plus random fluctuations of the dipolar axis (Volino et al. 1976) with $\langle \cos \theta \rangle = 0.75 (\bar{\theta} \approx 40°)$. Furthermore, the width of the quasi-elastic spectrum increases with Q, showing the presence of at least two Q-dependent components.

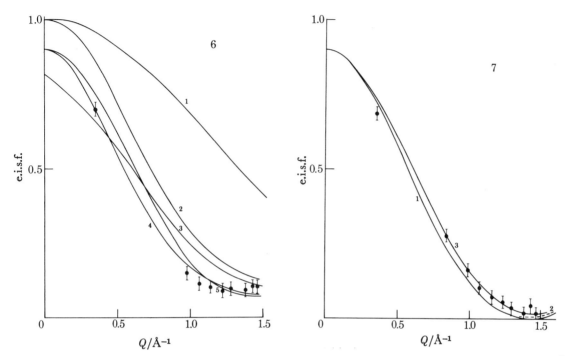

FIGURE 6. The elastic incoherent structure factor (e.i.s.f.) for t-butyl chloride at 202 K. The points are experimental values. The lines are the e.i.s.f. expected from the following models: (1) threefold jump reorientation of the methyl groups; (2) uniaxial rotational diffusion about the C–Cl axis; (3) as (2), but with the *maximum possible* correction for multiple scattering; (4) threefold jump reorientation about the C–Cl axis plus ca. 40° fluctuations of the long axis orientation and including the best estimate of multiple scattering effects; (5) uniaxial rotational diffusion plus a random lateral displacement of the molecule with $\langle x^2 \rangle^{\frac{1}{2}} \approx 0.7$ Å and including multiple scattering effects.

FIGURE 7. The elastic incoherent structure factor (e.i.s.f.) for the crystal I phase of t-butyl chloride at 236 K. ●, Experimental. The lines are calculated for the following models, including the effects of multiple scattering: (1) random rotational diffusion; (2) reorientations of C–Cl to point to all eight corners of a cube plus rotation around the C–Cl axis; (3) reorientations of C–Cl to point to the four corners of a tetrahedron plus rotation around the C–Cl axis (see text).

This can be fitted by a small step rotational motion as shown in figure 4, which would then imply that the additional displacements are on a similar time scale, or alternatively by a threefold rotation (perhaps in a softened potential; Dianoux & Volino 1977) together with the additional motion on a slightly different time scale. In either case, the data are consistent with uniaxial reorientation with a correlation time of about 4×10^{-12} s, and an additional motion is revealed which is either associated with a cooperative relaxation involving steric effects and changes of local symmetry, or overdamped fluctuations of the C–Cl axis which presage the molecular tumbling observed in the high temperature crystal I. We believe the latter to be the dominant motion because it is more consistent with the X-ray structural results and the small change of dielectric constant at the III–II transition.

The interpretation of the n.m.r. results must therefore be changed (and there seems to be no difficulty about this) so that the faster motion is the uniaxial whole-molecular rotation discussed above, while the slower motion must be the methyl reorientation. In this case the n.m.r. experiments suggest for the methyl reorientation $\tau/\mathrm{s} = (7 \pm 5) \times 10^{-12} \exp(11.7 \pm 0.9) \mathrm{~kJ~mol^{-1}}/RT$. From methyl torsion frequencies, Ratcliffe & Waddington (1976) estimate a barrier height $V = 18.5 \mathrm{~kJ~mol^{-1}}$; evaluating an average activation energy for reorientation gives $E \gtrsim 17 \mathrm{~kJ~mol^{-1}}$. Such discrepancies are, however, not uncommon and probably imply that the assumed potential is inadequate. The pre-exponential factor, on the other hand, is now reasonable, being of the order of the thermal free rotator correlation time (Brot 1969). Furthermore, an Arrhenius behaviour is expected for the methyl reorientation, with not much change through the transitions, because the barrier to reorientation is largely internal and therefore not strongly temperature dependent. This is certainly not true for motion about the C–Cl axis since, as we have demonstrated for pivalonitrile (Frost et al. 1980), and as the n.q.r. data of O'Reilly et al. (1973) suggest; the barrier hindering this motion is changing rapidly near the transitions. Consequently the activation energy and pre-exponential factor given by O'Reilly et al. for this motion (but assigned to methyl reorientation) are not very meaningful.

3.3. Crystal I

For crystal I the e.i.s.f. data (figure 7) are in reasonable agreement with the unequivocal quadrupole resonance, microwave absorption (see below) and p.m.r. linewidth results that the molecules are tumbling with at least cubic symmetry. Both the e.i.s.f. and $S_\mathrm{s}(Q, \omega)$ itself are fitted quite well by the simple model for rotational diffusion (equation (11)) with a best fit value for D_r^{-1} at 236 K of 8×10^{-12} s. The dielectric relaxation measurements give a value for the first spherical harmonic correlation time $\tau_1 = 5.4 \times 10^{-12}$ s (Brot & Lassier-Govers 1976) which may be compared with $(2D_\mathrm{r})^{-1} = 4 \times 10^{-12}$ s and the n.m.r. spin lattice relaxation measurements give the second spherical harmonic correlation time $\tau_2 = 3 \times 10^{-12}$ s which should be compared with $(6D_\mathrm{r})^{-1} = 1.3 \times 10^{-12}$ s. The agreement is fair, but together with analysis of dielectric (Lassier & Brot 1968) and X-ray data suggests that the tumbling is not entirely free but that weak minima occur in the orientational potential. Such a localization should give rise to a higher e.i.s.f., and indeed the experimental points are higher than expected for rotational diffusion: the difference is small but significant. The most likely orientations are those in which the site symmetry and the molecular symmetry have common elements, and Lassier & Brot have discussed the energies of the various possible orientations and the barriers between them.

The restricted range of Q and the single temperature for the present experiments does not permit a full analysis of the reorientations, but do enable us to identify the fastest motions. Thus the n.q.r. data, unit cell symmetry (f.c.c.) and entropy changes imply that on the time scale of the n.q.r. experiment (10^{-7} s), the C–Cl axis can point to at least the eight corners of the cubic cell (the molecular C_3 axis then coinciding with the crystal C_3 axis). The e.i.s.f. for this manifold is shown in figure 7 and is almost identical (for $Q \gtrsim 1.5 \mathrm{~\AA^{-1}}$) to that for random orientation, showing that the motion on the neutron time scale is more restricted than this. Since the rapid reorientation of the molecule about its C_3 axis must certainly occur in I as well as in II, perhaps the most likely additional rapid reorientation is rotation of the pseudo-tetrahedron about the C_3 axes of the methyl groups. The e.i.s.f. for this total motion, which moves the C–Cl axis among the four corners of the pseudo-tetrahedron, is in perfect agreement with the data when corrected for multiple scattering. This implies that the C_4 rotations of the pseudo-tetrahedron (and any other motions) are slower than $ca.$ 10^{-10} s. The data cannot distinguish whether the rotation about the C–Cl axis in I is threefold or by small steps, but it may be concluded that this motion is not more than about three times faster than the reorientation of the C–Cl axis among its four orientations, and the reorientation frequencies are in the range $1-5 \times 10^{11}$ rad s^{-1}.

3.4. Conclusion

The ability of i.n.q.e.s. to probe spacial as well as temporal aspects of the molecular motions has permitted unequivocal identification to be made of the basic type of reorientational motions taking place in the two disordered crystal phases of t-butyl chloride and determination of their correlation times. Differences from a previous neutron scattering study may be attributed to the better resolution of the present experiments. The results show clearly that whole molecule reorientation is more rapid than that of C_3 methyl group reorientation, in agreement with deductions from torsional frequencies, and show further that the two types of motion in II and I were wrongly identified in n.m.r. experiments.

4. Liquid crystals

4.1. Introduction

Thermotropic liquid crystals are intermediate phases that are often found between the normal solid and liquid phases of pure compounds. (A second type, lyotropics, are formed by surfactant molecules in concentrated solution, but these are not discussed here.) Liquid crystals are invariably made up of elongated, lath-like molecules, and as the temperature of a sample is increased the ordered solid phase changes to one or more different mesophases before the isotropic liquid is reached. Such behaviour is found in many organic compounds (see, for example, Gray 1979). A typical example, known as IBPBAC, has been chosen to illustrate how quasielastic neutron scattering has been used to study the mesophases. The molecular structure, chemical name and transition temperatures of IBPBAC are given below.

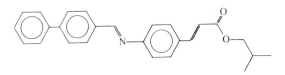

isobutyl 4-(4'-phenylbenzylidineamino)-cinnamate

crystal $\underset{\text{supercools}}{\overset{86\ °C}{\rightleftharpoons}}$ smectic E $\overset{114\ °C}{\rightleftharpoons}$ smectic B $\overset{162\ °C}{\rightleftharpoons}$ smectic A $\overset{206\ °C}{\rightleftharpoons}$ nematic $\overset{214\ °C}{\rightleftharpoons}$ isotropic liquid

Figure 8 shows schematically that the nematic phase has only orientational order of the long molecular axes, but the smectic phases are characterized by a layer structure. The different smectic phases are distinguished by the degree of orientational ordering about the long axes, the interlayer correlation and in some cases (but not IBPBAC) a tilt of the molecules with respect to the layers. X-ray diffraction experiments (see, for example, Doucet (1979) and Leadbetter *et al.* (1979a)) have shown that the packing is liquid-like in S_A layers, hexagonal on average in S_B layers and there is a herringbone packing of the lath-like molecules in the S_E layers. The director, n, is defined as the mean direction of the long axes in a liquid crystal. Monodomain samples may be prepared by carefully melting a single crystal or by cooling a magnetically aligned nematic phase.

FIGURE 8. Schematic view of the long molecular axis in the liquid crystalline phases of IBPBAC.

4.2. *Principle of the measurements*

The techniques used to study *t*-butyl chloride have also been used to elucidate the dynamics of several liquid crystal phases (Leadbetter *et al.* 1979b; Leadbetter & Richardson 1978, 1979). Incoherent quasielastic neutron scattering measurements have been used to observe and quantify the slow diffusive molecular motions of IBPBAC in its liquid crystalline phases, with the aim of achieving a better understanding of molecular interactions in these phases. A version of IBPBAC with deuterated isobutyl groups was used so that the results would be dominated by incoherent scattering from protons on the relatively rigid core of the molecule.

Great advantage can be taken from the fact that measurements can be done on aligned samples with Q parallel or perpendicular to n. For example, if the molecules are undergoing uniaxial rotational diffusion the scattering law is given by the formula

$$S(Q, \omega) = J_0^2(Qa \sin \theta) \, \delta(\omega) + \pi^{-1} \sum_{l=1}^{\infty} J_l^2(Qa \sin \theta) \frac{l^2 D_r}{(l^2 D_r^2)^2 + \omega^2},$$

where J_l is the lth cylindrical Bessel function, a is the radius of gyration of a proton, θ is the angle between Q and n, and D_r is the rotational diffusion constant.

Two extremes for the e.i.s.f. would be observable in an aligned sample:

$$\text{for} \quad Q \perp n, \text{ e.i.s.f.}_\perp = J_0^2(Qa);$$

$$\text{for} \quad Q \parallel n, \text{ e.i.s.f.}_\parallel = 1.$$

If the uniaxial rotational diffusion is restricted by some preferred orientations of the molecules, then the e.i.s.f.$_\perp$ will be higher than that given above (for $Qa < 3$).

The presence of an additional motion on the same timescale with an appreciable amplitude perpendicular to n, such as directional fluctuations of the long molecular axes, will tend to decrease the e.i.s.f.$_\perp$. Any diffusive motion parallel to n will result in the e.i.s.f.$_\parallel$ being reduced from unity. For instance, if the molecules undergo a random displacement parallel to n with a Gaussian 'infinite time' distribution about their mean positions, the e.i.s.f.$_\parallel$ would be given by the formula

$$\text{e.i.s.f.}_\parallel = \exp(-Q^2 \langle Z^2 \rangle), \tag{12}$$

where $\langle Z^2 \rangle^{\frac{1}{2}}$ is the r.m.s. amplitude of the displacement. These different motions can be distinguished by measurements on aligned liquid crystalline samples.

The above equations illustrate the general principle that for any localized motion, the scattering becomes entirely elastic as Q tends to zero:

$$S(Q, \omega) = \delta(\omega);$$

$$Q \to 0.$$

At sufficiently low Q, only translational diffusion will broaden the profile of the scattered neutrons. However, at low Q, the broadening from translational diffusion is very small and so very high resolution is necessary to observe it. There are therefore two types of experiment that have been done on liquid crystals:

(i) very high resolution measurements over a narrow energy transfer range at low Q, to give information about translational diffusion;

(ii) high resolution measurements over a wider energy transfer and Q range, to give information on the localized motions, such as rotation.

4.3. Translational diffusion

The IN10 high-resolution backscattering spectrometer at I.L.L., Grenoble, achieves a resolution of 1 µeV and covers an energy range of $|\hbar\omega| < 15$ µeV. In principle, translational diffusion constants parallel and perpendicular to $n(D_\parallel, D_\perp)$ can be obtained by measuring the Lorentzian broadening of the incident monoenergetic beam at $Q \ll 2\pi/d$ where d is the appropriate molecular dimension (cf. §2).

Figure 9 shows the diffusion constants of an aligned sample of IBPBAC in its liquid crystalline phases. Measurements were made at 0.14 Å$^{-1} < Q < 0.31$ Å and so the condition that $Q \ll 2\pi/d$ was well satisfied for $Q \perp n$ (where $d \approx 5$ Å, the molecular diameter) but not for $Q \parallel n$ (where $d \approx 27$ Å, the molecular length). The observed values of $D_\parallel$ in the S_A phase are therefore expected to be overestimates since the molecule is only followed for a distance that is comparable to its length. However, a detailed analysis (Richardson et al. 1980) has shown that the layers in the S_A only offer a weak barrier to diffusion and so the measured values of $D_\parallel$ are within 40 % of the correct values. The discontinuous jump in $D_\parallel$ at the smectic A to nematic transition reflects the disappearance of this barrier. In the smectic B phase the broadening was hardly observable with the available resolution and so the diffusion constants are best regarded as upper limits.

4.4. Localized molecular motions

The IN5 multichopper spectrometer (resolution typically 20 µeV) has been used to observe the quasielastic scattering from localized diffusive molecular motions, which, in liquid crystals, are more rapid than translational diffusion. Measurements have been made on IBPBAC in all three of its smectic phases and converted into experimental e.i.s.f. values. In the S_E and S_B phases the translational diffusion is too slow to broaden the instrumental resolution, but in the S_A both the elastic and quasielastic components were broadened by an amount ΔE ($\Delta E = 2\hbar D_t Q^2$) predicted by the translational diffusion constant measured on IN10. Figure 10 shows typical quasielastic scattering spectra from IBPBAC at 150 °C in the S_B phase. Surprisingly, there is more quasielastic scattering for $Q \parallel n$ than for $Q \perp n$, and the low e.i.s.f.$_\parallel$ (figure 11) indicates that there is a localized motion (on a timescale faster than 10^{-10} s) parallel to n (i.e. perpendicular to the smectic layers). The amplitude of the motion can be found by fitting the e.i.s.f.$_\parallel$ predicted by (12) to the observed e.i.s.f.$_\parallel$. The same motion has been found to increase smoothly from 1 Å at 100 °C in the S_E phase to 1.8 Å at 150 °C in the S_B, before dropping slightly to 1.3 Å in the S_A. The e.i.s.f.$_\perp$ in the S_B lies between that expected for uniaxial rotational diffusion and jump reorientation by π and so the molecules must undergo some form of reorientation between six (or more) sites (with $\tau \approx 3 \times 10^{-11}$ s), but with some sites preferred over others. In the S_E phase the e.i.s.f.$_\perp$ was found to be higher than that expected for reorientation by π and so the molecules can only be undergoing some overdamped libration about their long axes with an amplitude of *ca.* 40°. In the S_A phase, the e.i.s.f.$_\perp$ is low enough to be in good agreement with that for uniaxial rotational diffusion about the long molecular axes plus small-amplitude directional fluctuations of these axes. The width of the quasielastic part of the scattering implies a rotational diffusion constant, D_r of 4.7×10^{10} rad^2 s^{-1}.

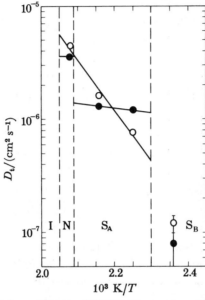

FIGURE 9. The translational diffusion constants of IBPBAC as measured by neutron scattering ($\bullet$, $D_\parallel$; $\circ$, $D_\perp$) as a function of inverse temperature. The lines serve to emphasize the discontinuity in $D_\parallel$ at the smectic A–nematic transition.

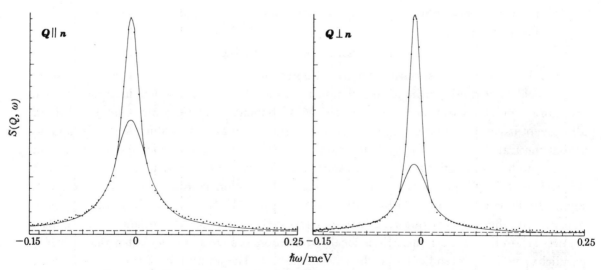

FIGURE 10. Quasielastic scattering from IBPBAC at 150 °C in the smectic B phase at $Q = 0.83$ Å^{-1} for $Q \perp n$ and $Q \parallel n$.

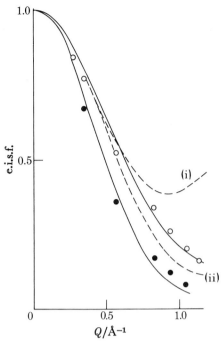

FIGURE 11. Elastic incoherent structure factors for IPBBAC. The experimental values for the smectic B phase at 150 °C are shown as points ($\bullet$, $Q \parallel n$; $\circ$, $Q \perp n$). The dashed lines are the e.i.s.fs expected for $Q \perp n$ if the molecules undergo (i) rotational diffusion or (ii) jump reorientation by π about their long axes. The solid lines are the result of fitting a model of restricted rotation plus a localized motion perpendicular to the layers in an imperfectly aligned sample to all the experimental data points.

4.5. Summary

These results from IBPBAC show how the dynamic rotational disorder about the long molecular axes changes along the S_E, S_B, S_A progression. Similar results from other substances have also extended this progression to the nematic phase. However, an important characteristic of all these phases appears to be that the molecules have the freedom to shuffle 1–2 Å parallel to their long axes (with $\tau < 10^{-10}$ s), and so any description of these phases entirely in terms of rotational disorder must be inadequate. The ability of molecules to move parallel to their long axes provides a mechanism by which some phase transitions take place (Leadbetter et al. 1979c). Certain liquid crystals have been found that undergo a smectic B to smectic H transition, not by a cooperative tilt of the molecules with respect to the layers but by a shift (at constant orientation) of 3–4 Å parallel to neighbouring molecules.

REFERENCES (Frost et al.)

Barnes, J. D. 1973 *J. chem. Phys.* **58**, 5193–5201.
Batley, M., Thomas, R. K., Heidemann, A., Overs, A. H. & White, J. W. 1977 *Molec. Phys.* **6**, 1771–1778.
Brot, C. 1969 *Chem. Phys. Lett.* **3**, 319–322.
Brot, C. & Lassier-Govers, B. 1976 *Ber. BunsenGes. phys. Chem.* **80**, 31–41.
Clough, S. & Heidemann, A. 1979 *J. Phys.* C **12**, 761–770.
Clough, S., Alefeld, B. & Suck, J. B. 1978 *Phys. Rev. Lett.* **41**, 124–128.
Clough, S., Heidemann, A. & Kraxenberger, H. 1979 *Phys. Rev. Lett.* **42**, 1298–1301.
Dianoux, A. J. & Volino, F. 1977 *Molec. Phys.* **34**, 1263–1277.
Dianoux, A. J., Volino, F. & Hervet, H. 1975 *Molec. Phys.* **30**, 1181–1194.
Doucet, J. 1979 In *The molecular physics of liquid crystals* (ed. G. R. Luckhurst & G. W. Gray), pp. 317–341. London: Academic Press.

Durig, J. R., Craven, S. M. & Bragin, J. 1969 *J. chem. Phys.* **51**, 5663–5673.
Frost, J. C. 1979 Ph.D. thesis, University of Exeter.
Frost, J. C., Leadbetter, A. J. & Richardson, R. M. 1980 *Faraday Discuss. chem. Soc.* **69**. (In the press.)
Goyal, P. S., Nawrocik, N., Urban, S., Domoslawski, J. & Natkaniec, I. 1974 *Acta phys. pol.* A **46**, 399–406.
Gray, G. W. 1979 In *The molecular physics of liquid crystals* (ed. G. R. Luckhurst & G. W. Gray), pp. 1–29. London: Academic Press.
Hüller, A. & Press, W. 1978 In *Neutron inelastic scattering*, vol. 1, pp. 231–253. Vienna: I.A.E.A.
Johnson, M. W. 1974 *A.E.R.E. Rep.* no. 7682.
Kushner, L. M., Crowe, R. W. & Smyth, C. P. 1950 *J. Am. chem. Soc.* **72**, 1091–1098.
Larsson, K. E., Mansson, T. & Olsson, L. G. 1978 In *Neutron inelastic scattering*, Vol. 2, pp. 435–469. Vienna: I.A.E.A.
Lassier, B. & Brot, C. 1968 *J. Chim. phys.* **65**, 1723–1732.
Leadbetter, A. J. & Lechner, R. E. 1979 In *The plastic crystalline state* (ed. J. N. Sherwood), pp. 285–320. London: John Wiley.
Leadbetter, A. J. & Richardson, R. M. 1978 *Molec. Phys.* **35**, 1191–1200.
Leadbetter, A. J. & Richardson, R. M. 1979 In *The molecular physics of liquid crystals* (ed. G. R. Luckhurst & G. W. Gray), pp. 451–483. London: Academic Press.
Leadbetter, A. J., Frost, J. C., Gaughan, J. P. & Mazid, M. A. 1979a *J. Phys., Paris* **40**, 185–192.
Leadbetter, A. J., Frost, J. C. & Richardson, R. M. 1979b *J. Phys., Paris* **40**, C3-125.
Leadbetter, A. J., Kelly, B. A., Mazid, M. A., Goodby, J. W. & Gray, G. W. 1979c *Phys. Rev. Lett.* **43**, 630–633.
Mansson, T. & Larsson, K. E. 1977 *J. chem. Phys.* **67**, 4996–5005.
O'Reilly, D. E., Peterson, E. M., Scheie, C. E. & Seyfarth, E. 1973 *J. chem. Phys.* **59**, 3576–3584.
Prager, M. & Alefeld, B. 1976 *J. chem. Phys.* **65**, 4927–4928.
Prager, M., Press, W., Alefeld, B. & Huller, A. 1977 *J. chem. Phys.* **67**, 5126–5132.
Press, W. & Prager, M. 1977 *J. chem. Phys.* **67**, 5752–5754.
Ratcliffe, C. I. & Waddington, T. C. 1976 *J. chem. Soc. Faraday Trans.* II **72**, 1821–1839.
Richardson, R. M., Leadbetter, A. J., Bonsor, D. & Kruger, G. J. 1980 *Molec. Phys.* (In the press.)
Rudman, R. & Post, B. 1968 *Molec. Cryst. liq. Cryst.* **5**, 95–110.
Schwartz, R., Post, B. & Frankuchen, I. 1951 *J. Am. chem. Soc.* **73**, 4490–4491.
Stiller, H. 1978 *J. molec. Struct.* **46**, 431–445.
Volino, F., Dianoux, A. J. & Hervet, H. 1976 *J. Phys., Paris* **37**, C355–364.

Percolation in antiferromagnetic insulators

By R. A. COWLEY, F.R.S.

Department of Physics, University of Edinburgh,
King's Buildings, Mayfield Road, Edinburgh EH9 3JZ, U.K.

The use of neutron scattering techniques to study problems in statistical mechanics is illustrated by describing recent experiments on magnetic–non-magnetic transition metal fluorides. The concentration of the magnetic ions is chosen to be close to the percolation point for the onset of long-range magnetic order. The results have confirmed that the percolation point is a multicritical point at which the long-range order may be destroyed by either geometrical or thermal disorder. The temperature scale of the thermal disorder is determined by the one-dimensional weak links in the large clusters. Results are obtained for the exponents in systems of dimensionality two and three, and with Ising and Heisenberg symmetry. In some systems the results agree with theoretical predictions, but there remain some discrepancies between theory and experiment and some aspects of the results which are not yet understood.

1. INTRODUCTION

One of the most rewarding applications of neutron scattering techniques during the last few years has been to study problems in statistical mechanics. Neutron scattering provides a uniquely detailed way of studying the phase transitions of magnetic systems, and by choice from the wide variety of magnetic systems it is possible to perform experiments on systems that closely approximate many of the simple models studied in statistical mechanics. Neutron scattering enables measurements to be made of the development of the long-range order below the phase transition, and the changes in the amplitude and correlation length of the magnetic fluctuations both above and below the transition temperature.

The study of the antiferromagnetic transition metal fluorides has proved to be particularly useful. The magnetic interactions are well known from low temperature spin wave measurements and are largely between only nearest neighbour magnetic ions; they can be of either Heisenberg character ($JS_1 \cdot S_2$ in Mn^{2+} salts) or Ising character ($JS_1^z S_2^z$ in Co^{2+} salts). Furthermore, systems can be found in which the magnetic ions are arranged in effectively one-dimensional chains, two-dimensional sheets or three-dimensional networks. In brief, classic experiments have now been performed on many of these systems. We note only the work of Birgeneau et al. (1971) on tetramethyl ammonium manganese chloride, which gave excellent agreement with Fisher's (1964) theory of linear chains of classical spins, the work of Ikeda & Hirakawa (1974) on K_2CoF_4, which agrees with Onsager's (1944) theory of the two-dimensional Ising model, and the work of Tucciarone et al. (1971) on $RbMnF_3$, which gave agreement with theories of the three-dimensional Heisenberg model. More recently this work has been extended to study more complex situations such as crossover phenomena and tricritical points.

The transition metal antiferromagnets have also proved to be excellent systems on which to study the effect of disorder on the excitations. Mixed single crystals can be grown in which the magnetic interactions are well known and of short range, while the effects of the disorder can be studied in detail by using neutron scattering techniques. The first experiments were studies of the low concentration limit, and demonstrated the existence of local spin modes near some impurities and resonant distortions of the spin wave spectrum near other impurities as reviewed by Cowley & Buyers (1972). More recently, work has been on more concentrated systems and has elucidated the nature of the excitations in highly disordered systems and the extent to which they can be described by theories based on the coherent potential approximation and by computer simulation techniques. The results of this have been reviewed (Cowley 1976; Cowley *et al.* 1979).

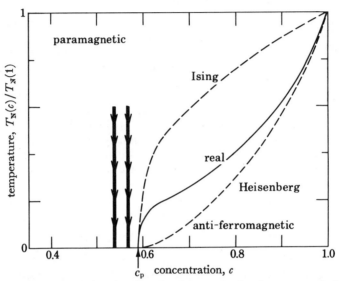

FIGURE 1. Schematic phase diagram of a two-dimensional square lattice close to percolation. The phase boundaries are sketched for Ising interactions, Heisenberg interactions and for the mixed interactions characteristic of $Rb_2Mn_cMg_{1-c}F_4$. The heavy vertical lines represent the locus of points at which the scattering was measured.

In this paper I wish to review some experiments where these two aspects of the usefulness of transition metal fluorides have been combined; namely, the study of the phase transitions of disordered systems, and in particular of systems close to the percolation limit in which sufficient non-magnetic ions have replaced the magnetic ions that the long-range magnetic order is destroyed. The phase diagram of this type of system is illustrated in figure 1.

The results described in this paper were obtained over the past few years at the Brookhaven National Laboratory, in collaboration with Dr R. J. Birgeneau and Dr G. Shirane. Full details of the experimental method and of the results are given in the full papers, which have now been prepared (Birgeneau *et al.* 1980; Cowley *et al.* 1980 *a*, *b*) and of which this report is a brief review.

The relevant aspects of percolation theory are reviewed in §2, together with a brief account of the neutron scattering from these systems. The experimental results are described in §3 and the conclusions of the study given in a final section.

2. Percolation theory and neutron scattering

When a magnetic system with nearest neighbour interactions is diluted with sufficient non-magnetic ions, the system breaks up into separate clusters at a well defined concentration, c_p; this is known as the percolation point, and a schematic phase diagram of a magnetic system as a function of concentration and temperature is shown in figure 1. The percolation point is believed to be an example of a multicritical point (Stauffer 1976; Stanley et al. 1976) in the sense that the long-range magnetic order for $c = c_p$ and $T = 0$ may be destroyed geometrically by changing c to less than c_p or thermally by raising the temperature. Much of the work on percolation has concentrated on the geometric aspects at $T = 0$, as reviewed by Essam (1972). Initially, the problem was studied by Monte Carlo computer methods and by series expansion, while more recently renormalization group theory has been applied to the Hamiltonian found by Kastelyn & Fortuin (1969), who mapped the percolation problem onto a lattice gas model.

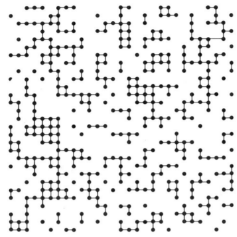

FIGURE 2. Clusters on a two-dimensional square lattice for $c = 0.50 < c_P = 0.59$.

There is therefore a correspondence between the properties close to the percolation point and the thermodynamic functions of a normal magnetic transition. In particular, the correlation length, ζ, is defined in terms of the pair connectedness, $S(r)$, which gives the probability that a site distance r away is in the same cluster as the site at the origin. Our experiments suggest that as with thermodynamic functions the Fourier transform of $S(r)$ is a Lorentzian.

$$S(q) = A/(\kappa^2 + q^2), \qquad (1)$$

where $\kappa = 1/\zeta$ is the inverse correlation length.

The mean size of the finite clusters plays the same role as the susceptibility at thermodynamic phase transitions and is given by $S_0 = S(0)$. Finally, the fraction of sites in the infinite cluster, $P(c)$, plays the same role as the magnetization at a thermodynamic phase transition.

Not surprisingly, the behaviour of these quantities as $c \to c_p$ is described by exponents. In particular, percolation theory suggests that for geometrically driven behaviour, $T = 0$ of figure 2, then

$$\kappa(c, T = 0) \approx (c_p - c)^{\nu_G}, \qquad (2a)$$

$$S_0(c, T = 0) \approx (c_p - c)^{-\gamma_G}, \qquad (2b)$$

$$P(c, T = 0) \approx (c - c_p)^{\beta_G}, \qquad (2c)$$

where the best values (Reynolds et al. 1977; Klein et al. 1978; Stanley 1977) of the exponents are listed in table 1.

To discuss the thermal critical behaviour it is first necessary to choose the appropriate temperature scale. In figure 2 I show a computer-generated two-dimensional square lattice with $c = 0.5$, which is less than $c_p = 0.59$. It is apparent that the large clusters consist of blobs joined by almost one-dimensional parts. These one-dimensional parts will dominate the thermal behaviour at low temperatures. Thus it is plausbile that the appropriate temperature scale (Stauffer 1976; Stanley et al. 1976; Birgeneau et al. 1976) is the inverse correlation length of the one-dimensional parts of the clusters, $\kappa_1(T)$. In the Ising and Heisenberg limits,

$$\kappa_1(T) \approx 2 \exp(-2J/kT) \quad \text{and} \quad \kappa_1(T) \approx kT/J, \quad \text{respectively.}$$

TABLE 1. EXPONENTS AND AMPLITUDES FOR PERCOLATION (Cowley et al. 1980b)

	ν_T	ν_G	γ_T	γ_G	D
2d Ising	1.32 ± 0.04	1.356 ± 0.015	2.40 ± 0.1	2.435 ± 0.045	0.96 ± 0.06
2d Heisenberg	0.90 ± 0.05	1.356 ± 0.015	1.50 ± 0.15	2.435 ± 0.045	1.00 ± 0.05
3d Ising	0.85 ± 0.10	0.845 ± 0.021	—	1.66 ± 0.07	—
3d Heisenberg	0.95 ± 0.04	0.845 ± 0.021	1.73 ± 0.15	1.66 ± 0.07	1.09 ± 0.03

Once the temperature scale has been established, we can introduce thermal exponents in analogy to (2) for $c = c_p$ and varying temperature by

$$\kappa(c_p, T) \approx (\kappa_1(T))^{\nu_T} \tag{3a}$$

and

$$S_0(c_p, T) \approx (\kappa_1(T))^{-\gamma_T}. \tag{3b}$$

Within the multicritical point picture, the exponents of the geometrically driven behaviour and the thermally driven behaviour are related by a crossover exponent, ϕ, so that $\gamma_T = \gamma_G/\phi$, and $\nu_T = \nu_G/\phi$. The exponent ϕ has been obtained as $\phi = 1$ for Ising systems by Wallace & Young (1978); more recently, Stinchcombe (1979) has suggested that this might be the case for Heisenberg systems as well.

Neutron scattering measurements give the magnetic spin–spin correlation function. The intensity for a wavevector transfer, Q, and integrated over all frequency transfers, ω, is determined by the static correlation function

$$I(Q) = \sum_{mn} \exp[iQ \cdot (R_m - R_n)] \langle S_m \cdot S_n \rangle, \tag{4}$$

where m and n denote the sites of the magnetic ions, and R_m and R_n and their positions. It is then apparent that $I(Q)$ is given in terms of the pair correlation function $S(q)$ of (1) when $I(Q) = S(q)\,\delta(Q-q-\tau) + N^2 P(c)^2\,\delta(Q-\tau)$, where τ is the nearest magnetic reciprocal lattice point to Q. These results show that neutron scattering may be used to determine A, κ and $P(c)$ and hence, in principle, the geometric and thermal exponents describing the percolation multicritical point.

3. EXPERIMENTAL RESULTS

The single crystals of mixed magnetic and non-magnetic transition metal fluorides were grown by Dr H. J. Guggenheim of Bell Laboratories and by the late Dr D. A. Jones of Aberdeen University. The samples were grown with concentrations close to c_p. Unfortunately, one of the difficult aspects of this work was determining the concentration and uniformity of the samples. Use was made of lattice parameter measurements, chemical analysis and the excitation spectra, but frequently the most accurate method was from the magnetic critical scattering at low temperatures and the use of percolation theory as described in the full accounts of this work. Unfortunately, the uncertainty in the concentration prohibited the experimental determination of the geometrical exponents. The experiments therefore consisted of measuring the temperature dependence of the correlations as a function of temperature for certain crystals or fixed concentrations as illustrated in figure 1.

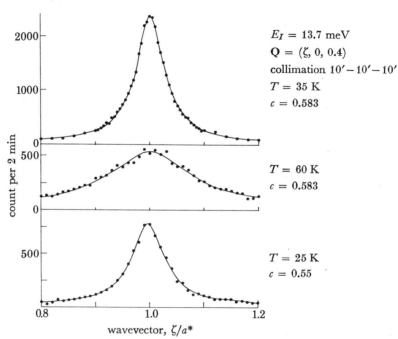

FIGURE 3. Neutron scattering distributions in $Rb_2Co_cMg_{1-c}F_4$ and fits to a two-dimensional Lorentzian, $S(q) = A/(q^2 + K^2) + B$ (———, fitted curve). (From Cowley et al. (1980a).)

The experiments were performed with a two-axis neutron spectrometer with the crystals mounted in a variable temperature cryostat. No evidence was found in any of the examples for chemical ordering of the different transition metal ions. The experiments were conducted with a sufficiently high incident neutron energy that the intensity gave a direct measure of the frequency integrated intensity, $I(Q)$.

Some of the results for the $Rb_2Co_cMg_{1-c}F_4$ system are illustrated in figure 3. The results show diffuse scattering, which decreases in width and increases in intensity as the system is cooled. The results for each temperature were fitted by assuming that the scattering is given by (1) and (4), convoluted by the experimental resolution function. A and κ were then determined to give the best fit, in a least-squares sense, to the experimental results. In each case a good fit was obtained, showing that the scattering is determined by a Lorentzian in reciprocal space so that the correlations decay exponentially in real space.

Initially, we discuss the results for $c < c_p$ when there is no long-range magnetic order. The temperature dependence of κ is shown in figure 4 for different concentrations in the two-dimensional Ising system, $Rb_2Co_cMg_{1-c}F_4$. In this system the exchange interactions are known from spin wave measurements (Ikeda & Hutchings 1978; Cowley et al. 1980a). The results suggest that the inverse correlation length is given by the simple form

$$\kappa = \kappa_G + \kappa_T, \qquad (5)$$

where κ_G depends solely on concentration and κ_T solely on the temperature. The thermal part is determined by the known one-dimensional correlation length, $\kappa_1(T)$, as

$$\kappa_T a = D(\kappa_1(T)\,a)^{\nu_T}, \qquad (6)$$

from which we can determine D and ν_T as given in table 1. Similarly, the exponent γ_T can be found from the intensity of the scattering. The results clearly confirm the multicritical point picture of percolation and (5) and (6), showing that the temperature scale is indeed determined by the one-dimensional weak links in the clusters. Furthermore, the measured exponents ν_T and γ_T are within error in agreement with the calculated geometrical exponents ν_G and γ_G (table 1), confirming that the crossover exponent, ϕ, is unity as obtained theoretically by Wallace & Young (1978).

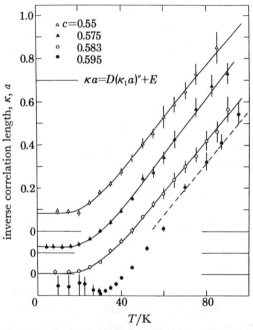

FIGURE 4. The temperature dependence of κ for four concentrations of $Rb_2Co_cMg_{1-c}F_4$. Note that for $c = 0.595$, long-range order occurs below 31 K. The solid lines show fits to (5) and (6). (Cowley et al. (1980a).)

Similar measurements were made on the analogous two-dimensional system, $Rb_2Mn_cMg_{1-c}F_4$, in which the magnetic interactions are predominantly of Heisenberg character (Cowley et al. 1977a). The results are illustrated in figure 5. The results show that the inverse correlation length decreases with decreasing temperature, until below about 4 K when the scattering becomes almost independent of temperature. This behaviour arises because at low temperatures the magnetic dipolar energy becomes comparable with kT. Consequently, although at high temperatures (20 K) the system is behaving like an isotropic Heisenberg system, on cooling the magnetic dipolar forces tend to align the spins along the unique c-axis and the system takes on Ising character. The inverse correlation length $\kappa_1(T)$ of an anisotropic linear chain was calculated by using a computer program written by Blume *et al.* (1975), and as shown in figure 5, equations (5) and (6) give a good account of the results. The value of ν_T is, however, quite different for this Heisenberg-like system from that for the Ising system discussed above. The value of the constant D is, in contrast, very similar.

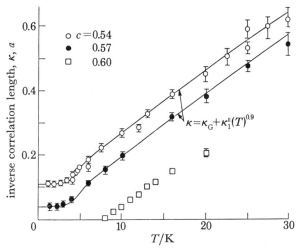

FIGURE 5. The inverse correlation length, κ, as a function of temperature for $Rb_2Mn_cMg_{1-c}F_4$. The crystal with $c = 0.60$ showed long-range order below 8.0 K. The solid lines show fits to (5) and (6). (Birgeneau *et al.* (1980).)

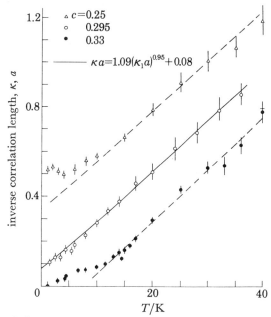

FIGURE 6. The inverse correlation length, κ, for three concentrations of $KMn_cZn_{1-c}F_3$. The solid line shows a fit to (5) and (6) the dotted lines are this fit with a different constant chosen to give agreement with other concentrations at high temperature. (Cowley *et al.* (1980b).)

Two three-dimensional systems have been studied. The $Mn_cZn_{1-c}F_2$ structure is tetragonal. In the pure system the magnetic interactions are of Heisenberg character but the magnetic dipolar forces produce a significant energy gap owing to the tetragonal structure (Nikotin *et al.* 1969). In the mixed system these forces give rise to a crossover to Ising behaviour at low temperatures, but the tetragonal structure makes it possible to determine the longitudinal and transverse scattering independently Consequently, it was shown (Cowley et al. 1977b) that both inverse correlation lengths were given by (5) and (6), if $\kappa_1(T)$ was replaced by the appropriate longitudinal and transverse inverse correlation lengths of the anisotropic linear chain. The parameters D and ν_T were the same for both longitudinal and transverse scattering.

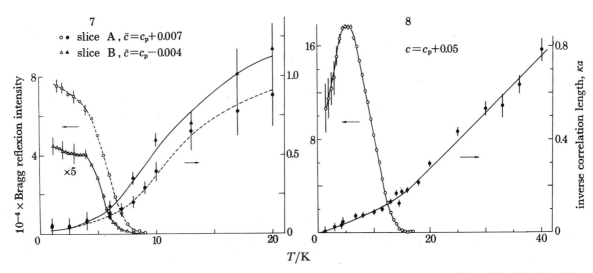

FIGURE 7. The Bragg scattering and the inverse longitudinal correlation length for two slices of the $Mn_cZn_{1-c}F_2$ system with concentrations as shown. Note that for neither slice is K zero at the onset of long-range order. (Cowley et al. (1977b).)

FIGURE 8. The Bragg scattering and inverse correlation length of $KMn_cZn_{1-c}F_3$ for $c > c_p$. (Cowley et al. (1980b).)

The other three-dimensional system is $KMn_cZn_{1-c}F_3$, which has the cubic perovskite structure. The exchange interactions in the pure material are of Heisenberg character between nearest neighbours and in the cubic symmetry there is no anisotropy gap in the spin wave spectrum owing to the magnetic dipolar forces. The results for the inverse correlation length in the mixed systems are shown in figure 6.

Once again, they are consistent with (5) and (6), and give results for the exponent ν_T and constant D as listed in table 1.

We now turn to discuss the results for $c > c_p$ when long-range order occurs at low temperatures. In the two-dimensional systems the behaviour for $c > c_p$ is at least qualitatively similar to expectations: the inverse correlation length decreases towards zero at the transition temperature, $T_N(c)$ (figures 4 and 5), and the long-range order as measured by the intensity of the magnetic Bragg reflexion increases monotonically with decreasing temperature. It is difficult to determine the exponents close to $T_N(c)$ because $T_N(c)$ changes rapidly with concentration c. Nevertheless, the indications are that the exponents are consistent with those occurring in pure K_2MnF_4 (Birgeneau et al. 1977) and pure K_2CoF_4 (Ikeda & Hirakawa 1974).

The results for $c > c_p$ for the two three-dimensional systems are considerably more difficult to understand. In neither $Mn_cZn_{1-c}F_2$ nor $KMn_cZn_{1-c}F_3$ is the measured critical scattering described by an inverse correlation length which decreases to zero at $T_N(c)$. Indeed, figures 7 and 8 show that κ is non-zero at $T_N(c)$ and continues to decrease as the temperature is lowered.

The scattering in these systems arises from three parts: the finite clusters, the backbone or multiply connected part of the infinite cluster, and the dangling ends of the infinite cluster. Presumably only the backbone orders at $T_N(c)$, and if this is a small enough part of the whole crystal then the critical scattering from the backbone might be obscured by the scattering from the finite clusters and the dangling ends. Although computer calculations (Kirkpatrick 1978) show that for a given $c - c_p$ the proportion of spins in the backbone is less in three dimensions than in two, the numbers suggest that it is surprising that no critical scattering was observed associated with $T_N(c)$.

The other surprising feature is the temperature dependence of the long-range order in $KMn_cZn_{1-c}F_3$ as shown in figure 8. The intensity of the Bragg reflexion increases on cooling from $T_N(c)$ ca. 13.5 to 6 K, but then decreases on further cooling. This result is difficult to understand. Our only suggestion is that 6 K is about the temperature at which the magnetic dipolar forces begin to play a role in these systems. Although these cancel in the cubic environment of pure $KMnF_3$, the random environments of the mixed crystal will lead to randomly directed dipolar forces. Possibly these destroy the long-range order as suggested by Aharony (1978), leading to the development of a spin glass phase at low temperatures.

4. Conclusions

Neutron scattering measurements have been made on a number of transition metal antiferromagnets close to the percolation concentration. The results have substantiated the multicritical point picture of percolation as being a competition between geometrically and thermally driven disorder. The temperature scale of the thermal disorder has been identified as being determined by the one-dimensional weak links in the large clusters. Measurements have been made of the thermal exponents γ_T and ν_T for various different systems. In Ising systems these are within error the same as the best estimates of the geometrical exponents as suggested by Wallace & Young (1978). In the Heisenberg systems the results differ from the geometrical exponents in contradiction to the suggestion of Stinchcombe (1979). Measurements have also been made of the amplitudes, but there are as yet no theoretical results with which to compare these results.

In the two-dimensional systems the onset of long-range order occurs in a manner qualitatively similar to that of the pure systems, but in the three-dimensional systems this is not the case. The critical scattering does not show the usual divergences as $T \to T_N(c)$, but is relatively smoothly varying near $T_N(c)$. This result and the reason for the behaviour of the long-range order at low temperatures in $KMn_cZn_{1-c}F_3$ (figure 8) must await further work. We hope that these results will stimulate further work on the multicritical point picture of percolation, so that it will be possible to compare the results with theory in more detail. Finally, it is clear that neutron scattering results have provided very significant and timely results for our understanding of the statistical mechanics of percolation phenomena, in addition to the earlier successes at other types of phase transition.

I am grateful to my collaborators on this project, Dr R. J. Birgeneau and Dr G. Shirane, and to the many others who have assisted with the experiments and with the understanding of their significance. Financial support at Brookhaven was provided by the Division of Basic

Energy Sciences, contract no. EY-76-C-02-0016, and at Edinburgh by the Science Research Council. The paper was written while I was a summer visitor at A.E.K. Risø, for whose hospitality I am grateful.

REFERENCES (Cowley)

Aharony, A. 1978 *Solid State Commun.* **28**, 667–670.
Birgeneau, R. J., Als-Nielsen, J. & Shirane, G. 1977 *Phys. Rev* B. **16**, 280–292.
Birgeneau, R. J., Cowley, R. A., Shirane, G. & Guggenheim, H. J. 1976 *Phys. Rev. Lett.* **37**, 940–943.
Birgeneau, R. J., Cowley, R. A., Shirane, G., Tarvin, J. A. & Guggenheim, H. J. 1980 *Phys. Rev.* B. **21**, 317–332.
Birgeneau, R. J., Dingle, R., Hutchings, M. T., Shirane, G. & Holt, S. L. 1971 *Phys. Rev. Lett.* **26**, 718–721.
Blume, M., Heller, P. & Lurie, N. A. 1975 *Phys. Rev.* B **11**, 4483–4497.
Cowley, R. A. 1976 *A.I.P. Conf. Proc.* **29**, 243–247.
Cowley, R. A. & Buyers, W. J. L. 1972 *Rev. mod. Phys.* **44**, 406–450.
Cowley, R. A., Birgeneau, R. J. & Shirane, G. 1979*a* In *Strongly fluctuating condensed matter systems* (Proc. NATO School in Geilo, Norway, April 1979) (ed. T. Riste), pp. 157–182. New York: Plenum Press.
Cowley, R. A., Birgeneau, R. J., Shirane, G., Guggenheim, H. J. & Ikeda, H. 1980*a* *Phys. Rev.* B **21**, 4038–4048.
Cowley, R. A., Shirane, G., Birgeneau, R. J. & Guggenheim, H. J. 1977*a* *Phys. Rev.* B **15**, 4292–4302.
Cowley, R. A., Shirane, G., Birgeneau, R. J. & Svensson, E. C. 1977*b* *Phys. Rev. Lett.* **39**, 894–897.
Cowley, R. A., Shirane, G., Birgeneau, R. J., Svensson, E. C. & Guggenheim, H. J. 1980*b* *Phys. Rev.* (In the press.)
Essam, J. W. 1972 In *Phase transitions and critical phenomena II* (ed. C. Domb & M. S. Green), pp. 197–270. New York: Academic Press.
Fisher, R. J. 1964 *Am. J. Phys.* **32**, 343–346.
Ikeda, H. & Hirakawa, K. 1974 *Solid State Commun.* **14**, 529–532.
Ikeda, H. & Hutchings, M. T. 1978 *J. Phys.* C. **11**, L529–532.
Kastelyn, P. W. & Fortuin, C. M. 1969 *J. phys. Soc. Japan Suppl.* **26**, 11.
Kirkpatrick, S. 1978 *A.I.P. Conf. Proc.* **40**, 99.
Klein, W., Stanley, H. E., Reynolds, P. J. & Coniglio, A. 1978 *Phys. Rev. Lett.* **41**, 1145–1148.
Nikotin, O. P., Lindgard, P. A. & Dietrich, O. W. 1969 *J. Phys.* C **2**, 1168–1173.
Onsager, L. 1944 *Phys. Rev.* **65**, 117–149.
Reynolds, P. J., Stanley, H. E. & Klein, W. 1977 *J. Phys.* A **10**, L203–209.
Stanley, H. E. 1977 *J. Phys.* A **10**, L211–220.
Stanley, H. E., Birgeneau, R. J., Reynolds, P. J. & Nicoll, J. F. 1976 *J. Phys.* C **9**, L553–560.
Stauffer, D. 1976 *Z. Phys.* B **22**, 161–171.
Stinchcombe, R. B. 1979 *J. Phys.* C **12**, 2625–2636.
Tucciarone, A., Lau, H. Y., Corliss, L. M., Delapalme, A. & Hastings, J. M. 1971 *Phys. Rev.* B **4**, 3206–3245.
Wallace, D. J. & Young, A. P. 1978 *Phys. Rev.* B **17**, 2384–2387.

Incommensurate structures

By J. D. Axe

*Department of Physics, Brookhaven National Laboratory,
Upton, New York 11973, U.S.A.*

A review is given of recent neutron scattering studies of displacive incommensurate structures, and the instabilities that occur within periodic crystalline structures that lead to their formation. The concept of a soft phonon, in some cases associated with an electronic screening anomaly, is useful but not always capable of a completely satisfactory description. An unusual one-dimensional liquid-like phase has been studied in the non-stoichiometric mercury compound $Hg_{3-\delta}AsF_6$.

Incommensurate structures

Incommensurate structures are peculiar quasi-crystalline substances that lack periodic translational symmetry not in a haphazard amorphous way but because two (or perhaps more) elements of translational symmetry are present which are mutually incompatible. Suppose $A(r)$ and $B(r)$ represent the spatial distribution of two characteristic properties of a material and that

$$A(r) = \sum_{\{G\}} A_G \, e^{iG \cdot r}; \quad B(r) = \sum_{\{G'\}} B_{G'} \, e^{iG \cdot r}. \tag{1}$$

The structure is incommensurate if the sets of reciprocal lattice vectors $\{G\}$ and $\{G'\}$ have only the trivial elements $G = G' = 0$ in common. Various cases are possible depending on what A and B represent, as shown in table 1.

Magnetic and compositional modulation are well known and will not be considered further, although in both cases neutron studies have contributed greatly to our present understanding of these phenomena. I shall instead concentrate on the latter two cases, the first of which consists of interpenetrating (or overlaid) lattices of different spacing. In contrast to the first three cases, displacive modulation involves a periodic displacement, say

$$u(r) = A \cos(q_0 \cdot r - \phi) \tag{2}$$

of the scattering centres away from an average position on a regular lattice site. It is not uncommon for the modulation amplitude, A, to disappear above a certain temperature, the material thereby transforming from the incommensurate structure to a commensurate one with the average structure. This review is largely concerned with neutron scattering studies of such displacive incommensurate phase transformations, and of what they reveal of the nature of incommensurate instabilities. A final section is concerned with the unusual behaviour of the quasi-one-dimensional intergrowth compound, $Hg_{3-\delta}AsF_6$.

Soft mode instabilities and charge density waves

The form of (2) suggests that we view an incommensurate phase transformation as a condensation or 'freezing-in' of a phonon with wavevector q_0, which might occur because the phonon frequency, $\omega^2(q_0)$, vanishes. Although introduced to explain ferroelectric ($q_0 = 0$)

transformations (Cochran 1960) and subsequently generalized to include other high symmetry wave vectors on the Brillouin zone boundary (Cochran *et al.* 1968), there are no fundamental restrictions on the wavevector of a soft phonon instability.

TABLE 1

incommensurate structure type	$A(r)$	$B(r)$	example
1. magnetic	magnetic density, M	nuclear density, ρ	Cr, r.e. metals
2. compositional	average density, $\langle \rho_1 + \rho_2 \rangle$	differential density, $\langle \rho_1 - \rho_2 \rangle$	CuAu II, feldspars
3. intergrowth overgrowth	lattice ρ_1	lattice ρ_2	Ar on graphite $Hg_{3-\delta}AsF_6$
4. displacive	displacement field, $u(r)$	average density, $\langle \rho \rangle$	quasi-1 D and 2D metals, others

On a more microscopic level, such an incommensurate transformation may result from a charge density wave (c.d.w.) instability (Overhauser 1968) in the conduction electrons near the Fermi surface of a metal. The occurrence of such an electronic instability requires a large electronic susceptibility $\chi_0(q_0)$, which is favoured when large portions of the Fermi surface are separated by the special wavevector q_0. This favourable nesting of the Fermi surface is much more probable for quasi-one- or two-dimensional metals where the Fermi surface becomes independent of some components of electron momenta. The c.d.w. instability is coupled to the phonons because of the screening effect of the conduction electrons on the 'bare' phonon frequencies, $\Omega(q)$. In the random phase approximation, the observed physical phonon frequencies $\omega(q)$ are roughly of the form

$$\omega^2(q) = \Omega^2(q) \left(\frac{1 - \lambda_q^2 \chi_0(q)}{1 - v_q \chi_0(q)} \right),$$

where λ_q is an electron–phonon coupling constant and v_q is a Fourier component of the electron–electron interaction. Theory thus predicts the c.d.w. instability to be accompanied by a soft phonon $\omega(q_0) \to 0$, where q_0 is determined by the Fermi surface nesting. The phonon softening can be viewed as a giant Kohn anomaly (Kohn 1959). Neutrons scatter not directly from the c.d.w. but rather from the nuclear distortions resulting from the condensed phonon mode. (From a theoretical viewpoint, a mixed spin and charge density wave appears possible, but no spin components have yet been detected in c.d.ws.)

C.D.ws in metals

In quasi-one-dimensional metals, the Fermi surface approaches a set of parallel planes orientated perpendicular to the one-dimensional axis and separated by $2k_F$. This is the most geometrically favourable situation for c.d.ws with nesting occurring over the whole Fermi surface. It is often called a Peierls transformation, since Peierls (1955) first studied it in the mean field approximation long before physical manifestations were known.

The most striking confirmation of the existence of giant Kohn anomalies occurs in the quasi-one-dimensional conductor KCP ($K_2Pt(CN)_4Br_x$). This is well represented in figure 1, which shows neutron scattering data for longitudinal acoustic phonons propagating along c^*, the one-dimensional Pt chain direction in this material (Carneiro *et al.* 1976). The anomaly is so sharply confined along c^* that a conventional representation in terms of a sharp phonon dispersion surface is not possible with experimental resolution. It is interesting to note that in spite of the unstable fluctuations shown in figure 1, it is now generally conceded that true incommensurate long-range order is not achieved in KCP, possibly because of the random potential from the non-stoichiometric ($x \approx 0.3$) Br ions. This is somewhat unfortunate, for this material is in other respects an example, *par excellence*, of Peierls's original idea.

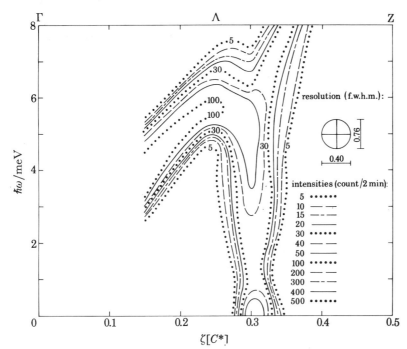

FIGURE 1. Intensity contours of inelastic scattering in KCP at 240 K showing extremely sharp Kohn anomaly near $q = 0.3\ c^*$. (After Carneiro *et al.* (1976).)

C.d.w. instabilities also occur in a number of quasi-two-dimensional transition metal chalcogenide compounds (Wilson *et al.* 1975). These materials have several different structural polytypes depending upon the stacking arrangement of the fundamental layers. I shall discuss only the 2H-polytypes of NbSe$_2$ and TaSe$_2$, for which the most extensive neutron studies are available. Unlike KCP, these substances undergo true phase transformations and their behaviour is at once more subtle and in some aspects better understood than is that of KCP.

Electron diffraction studies of both 2H-NbSe$_2$ and TaSe$_2$ showed satellite reflexions appearing at low temperature (Wilson *et al.* 1975). These peaks were originally thought to be commensurate with a spacing of one-third of the basal plane reciprocal lattice spacing a_H^*, but a subsequent neutron diffraction study (Moncton *et al.* 1975, 1977) showed that the ordering, which occurs at a well defined temperature, leads initially to incommensurate structures with wavevectors $q_\delta = \frac{1}{3}(1-\delta)\ a_H^*$ with $\delta \approx 0.02$ at the transformation temperatures ($T_0 = 33.4$ and 122.3 K for NbSe$_2$ and TaSe$_2$, respectively). The temperature dependence of the satellite intensities, which is proportional to the square of the c.d.w. amplitude, is shown in figure 2. The transformations appear to be continuous (i.e. second-order) although it is impossible to rule out the possibility of a small first-order discontinuity.

The most striking dynamical effect is a pronounced softening of the LA phonons propagating along a_H^*. As seen in figure 3, the wavevector minimum is very close to $\tfrac{1}{3}a_H^*$, as would be expected. This minimum is very insensitive to the c-axis component of momentum transfer, implying strong two-dimensional character. The square of the measured phonon frequencies at the minimum is plotted against temperature in figure 4. It is apparent that there is appreciable softening above T_0, and that this is reversed below T_0. It is also clear that the mode softening is far from complete at T_0, although the relatively poor experimental resolution may act to reduce the apparent sharpness of the anomaly. In addition to the inelastic scattering discussed above, there is also diffuse quasielastic critical scattering peaking near q_δ and at $T \approx T_0$.

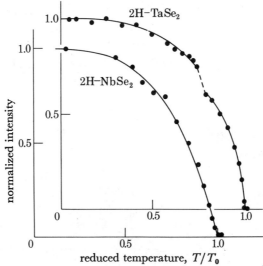

FIGURE 2. Temperature dependence of incommensurate satellite intensities in TaSe$_2$ and NbSe$_2$. (After Moncton et al. (1977).)

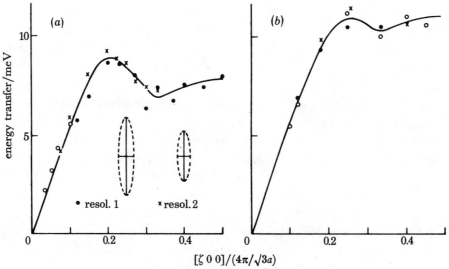

FIGURE 3. Phonon dispersion relations for Σ_1 phonon branches in TaSe$_2$ (a) and NbSe$_2$, (b) at 300 K. (After Moncton et al. (1977).)

An example of a nearly one-dimensional metal that does undergo a true c.d.w. transformation is afforded by TTF-TCNQ, an organic salt composed of tetrathiofulvalene (TTF$^+$) cations and tetracyanoquinodimethane (TCNQ$^-$) anions. Both are planar aromatic molecules that stack, plate-like, with overlapping partly filled π-orbitals responsible for the electronic conductivity. Incommensurate displacive modulations appear below $T_0 = 54$ K with a wavevector component along the stacking direction determined by the 1D Fermi surface, $2k_\text{F} = 0.295\,b^*$.

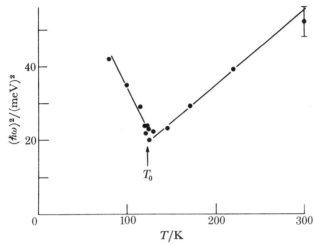

FIGURE 4. Temperature dependence of soft phonon energy in TaSe$_2$. T_0 marks the onset of c.d.w. formation. (After Moncton et al. (1977).)

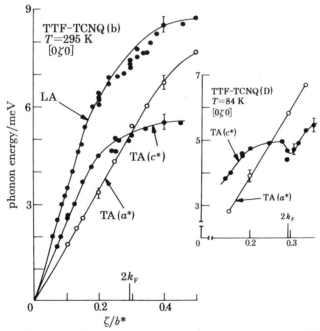

FIGURE 5. Dispersion of acoustic phonons propagating along b^* in TTF–TCNQ. (After Shapiro et al. (1977).)

Diffuse planes of strongly temperature-dependent X-ray scattering are observed below ca. 150 K, with the wavevector $2k_F = 0.295\ b^*$. Atomic displacements with components along both b^* (longitudinal) and c^* are observed. The inelastic neutron scattering data on deuterated samples (Shirane et al. 1976; Shapiro et al. 1977; Mook et al. 1977) show a distinct (but rather weak compared with KCP) dip in the TA branch with (mostly) c^*-axis polarization (see figure 5). At the minimum, the phonon energy decreases somewhat with decreasing temperature, but appears to remain more than 4 meV. If there is any anomalous dip in the LA branch at $2k_F$, it is considerably weaker than for TA(c^*), and the energies increase with increasing temperature. There is likewise no evidence for anomalies in the remaining TA(a^*) mode.

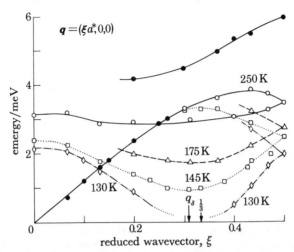

FIGURE 6. Dispersion of Σ_2, Σ_3 soft mode branch of K_2SeO_4 in an extended zone scheme. (After Iizumi et al. (1977).)

Insulators

In low dimensional metals the forces responsible for the instability are due to ion–electron–ion interaction and are long-ranged and oscillatory. The result, as we have seen, can be a sharp anomaly about $q_0 = 2k_F$. In insulators studied so far, the effective interaction ranges are shorter, leading to broader phonon anomalies, which are very conveniently studied by inelastic neutron scattering. In insulators, there is no essential requirement for incommensurate instabilities which favours low spatial dimensionalities, and indeed the known examples lack any obvious lower pseudo-dimensionality.

The most detailed neutron scattering study to date has been performed on K_2SeO_4 (Iizumi et al. 1977), which undergoes an incommensurate transformation at $T_0 = 130$ K. Figure 6 shows the dispersion of the soft phonon branch as a function of temperature. It is somewhat perverse that whereas all of the c.d.w. transformations in metals studied thus far do not follow the prediction of a simple soft mode instability, $\omega \to 0$ at $T = T_0$, such behaviour is seen in K_2SeO_4 for which no comparably simple and elegant microscopic description presents itself. On a phenomenological level, however, we can use the observed shape of dispersion relations such as that shown in figure 6 to deduce something about the effective force constants that couple planes of atoms perpendicular to the propagation direction. We find that from this point of view, the instability in K_2SeO_4 is brought about by an anomalously large force constant between planes of atoms which are third-nearest neighbours.

'LOCK-IN' TRANSFORMATIONS

We have seen that incommensurate structures arise from a competition of forces of varying range. However, even in the incommensurate state there are interactions at work which tend to restore periodicity. The ways in which these forces manifest themselves are illustrated in figure 7, which shows the temperature dependence of the wavevector of the incommensurate satellites in the transition metal dichalcogenides $NbSe_2$ and $TaSe_2$. The most striking feature is the abrupt change of the satellite vector in $TaSe_2$ from $q_1 = \frac{1}{3}(1-\delta)\,a_1^*$ to the commensurate value $\frac{1}{3}a_1^*$, which occurs at $T \approx 0.76\,T_0$. $NbSe_2$ does not achieve the $\frac{1}{3}a_1^*$ commensurate state even at the lowest attainable temperatures, but as in $TaSe_2$ the satellite wavevector shows a pronounced temperature dependence, whose origin is closely related to the 'lock-in' phenomenon itself.

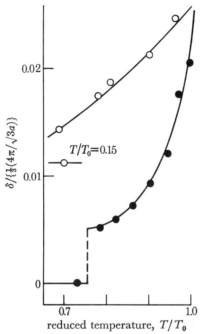

FIGURE 7. Temperature dependence of the satellite wavevector $q_\delta = \frac{1}{3}(1-\delta)\,a^*$ in $TaSe_2$ (●) and $NbSe_2$ (○). (After Moncton et al. (1977).)

Similar lock-in transformations have also been observed in other incommensurate systems. It would be inappropriate to present an extended discussion of these interesting transformations here, although neutron scattering observations were responsible both for their discovery and initial elucidation (Moncton et al. 1977). The key point is to recognize that purely sinusoidal modulation (equation (2) with ϕ constant) cannot take advantage of the periodic potential of the average lattice. The regions where the displacements are in phase with the potential are exactly cancelled by equally large and numerous out-of-phase regions. However, with the proper spatial variation of $\phi(r)$ the in-phase regions grow at the expense of the out-of-phase regions, thus lowering the total energy. A careful analysis of this effect (McMillan 1976) shows that this also produces a gradual pulling of the wavevector q_0 away from its initial (zero amplitude) value toward one commensurate with that of the average lattice. This phase modulation also produces additional secondary satellite reflexions that are initially weak but which grow to intensities comparable to the primary satellites near lock-in. Such behaviour has been observed in $TaSe_2$.

Phasons

What happens if we extend the soft mode picture of the phonons just above T_0 to discuss the lattice dynamics of the incommensurate state? A simple analysis assuming purely sinusoidal static displacements (McMillan 1975; Axe 1976) produces results shown schematically in figure 8. Above T_0, there is a soft branch with a minimum frequency $\omega(\boldsymbol{q}_0) \to 0$ as $T \to T_0$.

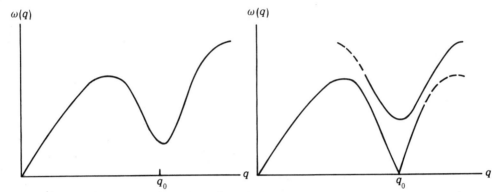

FIGURE 8. Schematic illustration of the dispersion relation for a material undergoing an incommensurate displacive phase transformation: (a) above T_0 there is a soft branch with a minimum at q_0; (b) below T_0 there is a 'splitting' of the modes into a gapless 'phase' branch and an upper 'amplitude' branch.

Below T_0 the added presence of the static displacements require new harmonic modes to be constructed from linear combinations of phonon wavevectors $(\boldsymbol{q}+\boldsymbol{q}_0)$ and $(\boldsymbol{q}-\boldsymbol{q}_0)$ respectively. This results in a splitting of the modes into an upper 'branch' for which

$$u_+(\boldsymbol{r}, t) \approx \cos(\boldsymbol{q}_0 \cdot l - \phi)\, e^{i(\boldsymbol{q}\cdot\boldsymbol{r} - \omega_+ t)} \tag{3}$$

and a lower 'branch' for which

$$u_-(\boldsymbol{r}, t) \approx \sin(\boldsymbol{q}_0 \cdot \boldsymbol{r} - \phi)\, e^{i(\boldsymbol{q}\cdot\boldsymbol{r} - \omega_- t)} \tag{4}$$

Comparison of (2)–(4) shows that for small q the upper branch is equivalent to a time-dependent modulation of the *amplitude* of the static displacements, whereas the lower branch represents a modulation of their *phase*. Inevitably, the term 'phason' has become accepted nomenclature for these latter modes, which are gapless and exhibit linear acoustic-like dispersion, $\omega_-(q) = vq$. However, the velocity v has no relation to the velocity of sound and the phasons are not to be confused with acoustic phonons, as figure 8 makes clear.

Of course, I have already pointed out that the static displacements are in general more complex than (2) suggests. Furthermore, we have seen that incomplete phonon softening renders the whole soft mode picture of little more than qualitative value in many cases. Nevertheless, we know on more general grounds that the absolute phase of the static distortion is not fixed by energetic considerations in an incommensurate structure and this fact alone guarantees a gapless phason branch. (Technically, phasons are Goldstone modes associated with a broken continuous symmetry. Only a pinning of the overall phase of the static distortion by lock-in terms or impurities will cause a gap to appear in the phason spectrum.) Fröhlich (1954) first recognized the unusual character of these excitations and used them to construct a novel and as yet unobserved mechanism for superconductivity.

In view of the unusual nature of phason excitations, the possibility that they make interesting contributions to various thermal and transport properties and the fact that inelastic neutron scattering seems ideally suited to their study, it is curious (and somewhat disconcerting) to note how little has in fact been learned, through no lack of effort. In fact, I am currently unaware of any *convincing* direct evidence for the existence of low-frequency propagating phason modes in any of the displacive incommensurate phases so far studied, although there is commonly observed unresolved additional scattering, which may represent *very* low frequency propagating and/or overdamped excitations. Fresh experiments and ideas are needed.

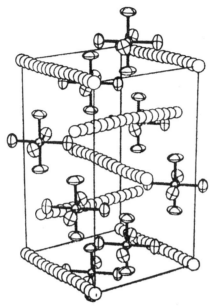

FIGURE 9. The structure of $Hg_{3-\delta}AsF_6$. (After Shultz *et al.* (1978).)

$Hg_{3-\delta}AsF_6$: A ONE-DIMENSIONAL LIQUID

This remarkable material, whose structure is shown in figure 9 consists of a tetragonal AsF_6 lattice in which there are open non-intersecting channels running parallel to both basal plane edges. These channels are filled with tightly packed chains of Hg ions. The observed interchain Hg–Hg distance is such that $3 - \delta$ ($\delta \approx 0.18$) Hg ions can be accommodated within a unit cell dimension of the AsF_6 lattice (Schultz *et al.* 1978). The low-temperature diffraction pattern consists of two distinct series of Bragg reflexions from separate, well ordered and incommensurate sublattices. The material is thus an example of an incommensurate structure of the intergrowth type (table 1). (In these materials, the Goldstone phason modes *are* independent acoustic phonons propagating on their respective sublattices and they *have* been directly observed by neutron scattering (Hastings *et al.* 1977; Heilmann *et al.* (1978)).

Above 120 K, the Bragg peaks of the Hg lattice disappear, to be replaced by a series of narrow sheets of scattering perpendicular to and spaced at a regular interval along the Hg chain axes. Such a one-dimensional diffraction pattern shows that the position along the chain of an arbitrarily chosen origin Hg ion is no longer fixed either with respect to the AsF_6 lattice or with respect to neighbouring Hg chains. Although the sheets of scattering were initially thought of as elastic 'Bragg sheets', this notion violates very general theorems that show the impossibility of true long-range periodic order in one dimension. Emery & Axe (1978) analysed the behaviour of a one-dimensional chain of atoms bound by harmonic nearest neighbour forces. In this model, sheets of scattering have a typically liquid-like diffraction pattern, as shown in figure 10, although in this figure the parameter σ/d, which characterizes the mean square thermal fluctuations relative to the mean near-neighbour spacing, d, is for illustrative purposes chosen to be considerably larger than is appropriate for $Hg_{3-\delta}AsF_6 (\sigma/d \approx 2.5 \times 10^{-2}$ at 300 K.)

For small (σ/d) the peak profiles are predicted to be Lorentzian, the width of successive peaks, $n = 1, 2, \ldots$, being proportional to n^2, the proportionality constant expressible in terms

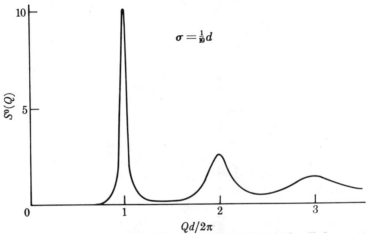

FIGURE 10. The diffraction pattern of a one-dimensional harmonic liquid. For small σ/d the width of successive peaks is proportional to n^2.

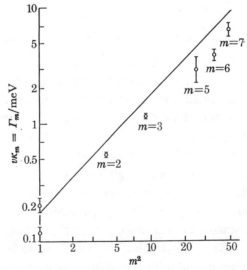

FIGURE 11. A comparison of experimental and theoretical values of the successive peak widths in $Hg_{3-\delta}AsF_6$ at 300 K. (After Heilmann et al. (1979).)

of the velocity of sound along the chain, v_1. Experiments measuring the width of successive sheets up to $n = 7$ were subsequently performed (Heilmann *et al.* 1979). The results, shown in figure 11, not only confirm the predicted n^2 dependence, but the value of v_1 derived from the measured widths is within 20% of the directly measured values (Heilmann *et al.* 1978). Thus the Hg chains do, at high temperature, behave like independent columns of one-dimensional harmonic liquid.

This review is the by-product of collaborations and discussions with many colleagues, especially V. J. Emery, M. Iizumi, D. E. Moncton and G. Shirane, to whom I am very grateful.

Research was supported by the Division of Basic Energy Sciences, U.S. Department of Energy, under contract no. EY-76-C-02-0016.

References (Axe)

Axe, J. D. 1976 In *Proceedings of Gatlinburg Neutron Scattering Conference* (ed. R. M. Moon), pp. 353–378. E.R.D.A.
Carneiro, K., Shirane, G., Werner, S. A. & Kaiser, S. 1976 *Phys. Rev.* B **13**, 4258–4263.
Cochran, W. 1960 *Adv. Phys.* **9**, 387–412.
Cochran, W. & Zia, A. 1968 *Physica Status Solidi* **25**, 273–274.
Emery, V. J. & Axe, J. D. 1978 *Phys. Rev. Lett.* **40**, 1507–1511.
Fröhlich, H. 1954 *Proc. R. Soc. Lond.* A **223**, 296–307.
Hastings, J. M., Pouget, J. P., Shirane, G., Heeger, A. J., Miro, N. D. & MacDiarmid, A. G. 1977 *Phys. Rev. Lett.* **39**, 1484–1488.
Heilmann, I. U., Hastings, J. M., Shirane, G., Heeger, A. J. & MacDiarmid, A. G. 1978 *Solid State Commun.* **29**, 469–473.
Heilmann, I. U., Axe, J. D., Hastings, J. M., Shirane, G., Heeger, A. J. & MacDiarmid, A. G. 1979 *Phys. Rev.* B **20**, 751–762.
Iizumi, M., Axe, J. D., Shirane, G. & Shimaoka, K. 1977 *Phys. Rev.* B **15**, 4392–4400.
Kohn, W. 1959 *Phys. Rev. Lett.* **2**, 393–396.
McMillan, W. L. 1975 *Phys. Rev.* B **12**, 1187–1194.
McMillan, W. L. 1976 *Phys. Rev.* B **14**, 1496–1508.
Moncton, D. E., Axe, J. D. & DiSalvo, F. J. 1975 *Phys. Rev. Lett.* **34**, 734–736.
Moncton, D. E., Axe, J. D. & DiSalvo, F. J. 1977 *Phys. Rev.* B **16**, 801–810.
Mook, H. A., Shirane, G. & Shapiro, S. M. 1977 *Phys. Rev.* B **16**, 5233–5237.
Overhauser, A. W. 1968 *Phys. Rev.* **167**, 691–698.
Peierls, R. F. 1955 *Quantum Theory of Solids*. Oxford: Clarendon Press.
Shapiro, S. M., Shirane, G., Garito, A. F. & Heeger, A. J. 1977 *Phys. Rev.* B **15**, 2413–2418.
Shirane, G., Shapiro, S. M., Comes, R., Garito, A. F. & Heeger, A. J. 1976 *Phys. Rev.* B **14**, 2325–2331.
Shultz, A. J., Williams, J. M., Miro, N. D., MacDiarmid, A. G. & Heeger, A. J. 1978 *Inorg. Chem.* **17**, 646–652.
Wilson, J. A., DiSalvo, F. J. & Mahajan, S. 1975 *Adv. Phys.* **24**, 117–159.

Elementary excitations in liquid ^{3}He

By K. Sköld†‡ and C. A. Pelizzari‡

† The Studsvik Science Research Laboratory, S-61182 Nyköping, Sweden
‡ Solid State Science Division, Argonne National Laboratory, Argonne, Illinois 60439, U.S.A.

The two isotopes of helium offer unique opportunities to test fundamental theories of quantum fluids. In liquid ^{4}He, a Bose liquid, the elementary excitations have been studied extensively over the last two decades by neutron inelastic scattering. Similar studies of liquid ^{3}He, the Fermi liquid counterpart of liquid ^{4}He, have become possible only recently. From the results obtained so far for normal liquid ^{3}He, the existence of zero sound as a well defined density excitation at finite wavevectors ($q \approx q_F$) and finite temperatures ($T \approx T_F$) has been verified and the spin fluctuation spectrum has been measured. By comparing the neutron scattering results for the density fluctuation spectrum with R.P.A. calculations with generalized polarization potentials, important information has been obtained about the effective spin-symmetric interaction between ^{3}He atoms. The spin fluctuation spectrum is in agreement with the paramagnon model if a value close to unity is assumed for the contact dimensionless paramagnon parameter $\bar{I}$. This implies that the nuclear spin system in normal liquid ^{3}He is close to ferromagnetic.

Introduction

A wealth of information regarding the elementary excitation spectrum in liquid ^{4}He has been obtained from neutron inelastic scattering measurements over the past 20 years (Price 1978). Practically all the information that is now available about the density fluctuation spectrum on a microscopic level has been obtained from studies of this kind. In liquid ^{3}He it is only recently that such measurements have become possible. The reason for this is the large neutron absorption cross section of the ^{3}He nucleus, which renders neutron scattering experiments rather difficult in this system.

In order to specify completely the dynamics of liquid ^{3}He, it is necessary to determine both the dynamics of the atoms (density fluctuations) and of the nuclear spin system (spin fluctuations). It is a most fortunate circumstance that the neutron probe couples both to the density fluctuations and to the spin fluctuations. This is due to the fact that the neutron scattering amplitude of the ^{3}He nucleus is different for the singlet and the triplet scattering states. The neutron scattering function therefore contains two components, namely the spin fluctuation scattering and the density fluctuation scattering.

The first neutron scattering results for liquid ^{3}He are those reported by Scherm et al. (1974). More extensive results were later reported by Stirling et al. (1975, 1976). These measurements were all made at the high-flux reactor at Institut Laue–Langevin in Grenoble. The spectra observed by the Grenoble group show a broad peak of inelastically scattered neutrons around energy transfers corresponding to the excitation of single particle–hole pairs, i.e. single particle excitations of the kind expected for a Fermi fluid, plus a long tail at larger energies. No evidence of a collective density mode of the zero sound type, which was predicted over 10 years ago by

Pines (1966), was observed in these experiments. The lowest temperature of the Grenoble experiments was 0.63 K and spectra were recorded for wavevector transfers larger than about 1 Å^{-1}†. In a subsequent experiment by the present authors (Sköld et al. 1976), the neutron scattering function for liquid ^{3}He at 15 mK was measured for wavevectors in the range 0.8 Å^{-1} ⩽ q ⩽ 2.2 Å^{-1}. These measurements were made at the CP-5 reactor at Argonne National Laboratory and the results show structure not observed in the Grenoble results. In particular, for q ⩽ 1.3 Å^{-1} the spectra display two peaks, namely one at ca. 0.2 meV and one at ca. 1 meV. The low energy peak is within the particle–hole band and is identified as spin-fluctuation scattering. The peak at ca. 1 meV is the zero sound mode predicted by Pines (1966). In a second experiment, the Argonne Group measured the scattering function for wavevectors down to 0.55 Å^{-1} (Sköld & Pelizzari 1977). In this case, dispersion of the zero sound mode is observed at the smaller wavevectors. In a recent experiment, the Argonne group has determined the scattering function at 40 mK and at 1.2 K and for wavevectors in the range 0.4 Å^{-1} ⩽ q ⩽ 2.2 Å^{-1} (Sköld & Pelizzari 1978a). These experiments were designed to study the temperature dependence of the scattering function and, in particular, to elucidate the discrepancy of the results obtained by the Argonne group and those obtained by the Grenoble group. The results show that for q ≲ 1.2 Å^{-1} and T = 1.2 K the zero sound mode is still a well defined excitation but that the spin fluctuation peak shows considerable broadening at the higher temperature. Thus, the discrepancy of the results obtained by the two groups can not be reconciled by the difference in the temperatures at which the results were obtained.

In the present report the results obtained by the Argonne group are discussed in some detail. The neutron scattering results are compared with theoretical predictions obtained from R.P.A. calculations with generalized polarization potentials in the density fluctuation spectrum and with the predictions of the paramagnon model in the spin fluctuation spectrum. It is shown that the density fluctuation spectrum can be fitted with the R.P.A. results if q-dependent effective interactions and a q-dependent effective mass is assumed. The experimental results for the spin fluctuation spectrum are in good agreement with the paramagnon model with the paramagnon parameter close to 1, i.e. the nuclear spin system appears to be close to ferromagnetic.

Theoretical overview

The theoretical models considered in this paper all have the same basic structure, namely that of the R.P.A. theory. In this case the dynamical function of interest is the density–density (spin–spin) response function $\chi^{C,I}(q, E)$, where C labels the density function and I labels the spin function. The corresponding neutron scattering functions are related to the response functions by

$$S^{C,I}(q, E) = -\pi^{-1} \operatorname{Im} \chi^{C,I}(q, E). \tag{1}$$

In the R.P.A. theory the response function is expressed in terms of the screened response function, $\chi_{sc}^{C,I}(q, E)$, and the effective interaction, $\psi^{C,I}(q)$:

$$\chi^{C,I}(q, E) = \frac{\chi_{sc}^{C,I}(q, E)}{1 + \psi^{C,I}(q) \chi_{sc}^{C,I}(q, E)}. \tag{2}$$

The screened response function describes the response of the system to the sum of an applied external field and the polarization fields produced by the density (spin) fluctuations. The effective interaction measures the strength of the polarization fields.

† 1 Å = 10^{-10} m = 10^{-1} nm.

In the approach taken by Aldrich & Pines (1978), in addition to the scalar polarization field produced by density fluctuations, the vector polarization field produced by current fluctuations is also considered. The scalar interaction is derived from the Fourier transform of an assumed analytical form for the effective two-body potential in r-space:

$$f_q^s = \frac{4\pi}{V} \int_0^\infty dr\, r^2 \frac{\sin qr}{qr} \tfrac{1}{2}\{f^{\uparrow\uparrow}(r) + f^{\uparrow\downarrow}(r)\} \qquad (3)$$

and

$$f_q^a = \frac{4\pi}{V} \int_0^\infty dr\, r^2 \frac{\sin qr}{qr} \tfrac{1}{2}\{f^{\uparrow\uparrow}(r) - f^{\uparrow\downarrow}(r)\}, \qquad (4)$$

where $f^{\uparrow\uparrow}(r)$ ($f^{\uparrow\downarrow}(r)$) is the interaction between particles of parallel (antiparallel) spins. The spin symmetric (s) and the spin asymmetric (a) interactions apply to the density fluctuation and the spin fluctuation fields respectively. The repulsive part of the effective interaction is parametrized in the following form:

$$f^{\uparrow\uparrow}(r) = a^{\uparrow\uparrow}[1 - (r/r_c^{\uparrow\uparrow})^8] \qquad r \leqslant r_c, \qquad (5)$$

with a similar expression for the spin asymmetric interaction. For $r > r_c$, an attractive van der Waals interaction is assumed. For f_q^s it is further postulated that

$$\tfrac{1}{2}\{f^{\uparrow\uparrow}(r) + f^{\downarrow\uparrow}(r)\} = a_s[1 - (r/r_s)^8] \qquad r \leqslant r_s, \qquad (6)$$

i.e. the same functional form as in (5) is assumed. The value of a_s is determined by $f_q^s = 0$ from the corresponding Landau parameter, while r_s is determined from a comparison with the experimental neutron scattering results. In the spin asymmetric interaction, we consider the effect of differences in $r_c^{\uparrow\uparrow}$ and $r_c^{\uparrow\downarrow}$ on $S^{\text{I}}(q, E)$. Aldrich & Pines (1978) also include the vector interaction $(\omega/q)^2 g_q^s$ in their description of the density fluctuation spectrum. The parameter g_q^s is related to the effective mass through

$$m_q^* = m_0 + N g_q^s, \qquad (7)$$

where m_0 is the bare ^{3}He mass and N is the number density. The effective mass parameter m_q^* (or g_q^s) is a phenomenological parameter in the theory and is adjusted to fit the neutron scattering data. Finally, to allow for multi particle–hole contributions to χ_{sc}, the one-particle χ_{sc} is renormalized according to

$$\omega_{\text{sc}}^{\text{C, I}}(q, E) = \alpha_q^{\text{C, I}} \chi_0^*(q, E) + \chi_{\text{m}}^{\text{C, I}}(q, E), \qquad (8)$$

where χ_0^* is the Lindhard response function with effective mass from (7), α_q is another phenomenological parameter and χ_{m} is approximated by

$$\chi_{\text{m}}^{\text{C, I}}(q, E) = \chi_{\text{m}}^{\text{C, I}}(q, 0). \qquad (9)$$

The phenomenological constants that appear in the theory are chosen such that in the limit $q \to 0$ the Landau–Fermi liquid theory is recovered.

The R.P.A. results with the phenomenological interactions described above are compared with the neutron scattering results for the density fluctuation spectrum. As discussed by Aldrich & Pines (1978), the same approach can also be used to analyse the spin fluctuation spectrum. In this case the effective interaction is due to subtle differences between the interaction of particles of parallel spins and the interaction of particles of antiparallel spins (compare equation (4)). It is observed that a difference of only 3 % in the range of the repulsive core could lead to a spin-wave instability at finite q.

In a recent publication by Beal-Monod (1979), the spin fluctuation spectrum is analysed within the framework of the R.P.A. theory but with the parameters chosen in accordance with the values appropriate to the paramagnon model (Doniach & Engelsberg 1966). In this case, the tendency to magnetic ordering is emphasized, the mass enhancement is neglected and the interaction is represented by the contact spin–spin repulsion parameter I:

$$\chi^{\mathrm{I}}(q, E) = \chi_0(q, E)/\{1 + I\chi_0(q, E)\}, \tag{10}$$

where χ_0 is the Lindhard function for the bare mass. From the neutron scattering results published earlier by the present authors (Sköld et al. 1976), Beal-Monod obtains $I \approx 0.9$. This is consistent with the experimentally observed enhancement in the static susceptibility. As $I = 1$ is the Stoner criterion for a ferromagnetic instability, the results imply that normal liquid ^{3}He at low temperatures ($T \ll T_{\mathrm{F}}$) is nearly ferromagnetic. These conclusions are supported by the results presented in this paper in which recent and more extensive neutron scattering results are compared with the predictions of the paramagnon model.

Experimental arrangements

The experimental arrangements are described in detail by Sköld et al. (1980). In the present context a brief summary of the most important aspects will therefore suffice.

The experiments reported here present several difficulties not ordinarily encountered in neutron scattering measurements. Most importantly, the large ratio of the neutron absorption cross section and the neutron scattering cross section of the ^{3}He nucleus ($\sigma_{\mathrm{a}} \approx 11\,000$ b† at $\lambda_{\mathrm{n}} \approx 4$ Å; $\sigma_{\mathrm{s}} \approx 6$ b) severely limits the statistical accuracy of the experimental results. Also, the weak signal from the sample requires that the experimental arrangement be designed such that systematic errors are minimized as much as possible. To study the fully developed quantum statistical properties of a Fermi fluid the temperature should be well below the Fermi temperature. For liquid ^{3}He ($T_{\mathrm{F}}^* \approx 1.5$ K) this implies that the temperature of the measurement should be less than a few tenths of a kelvin and a dilution cryostat is therefore required.

The neutron scattering experiments were made at the TNTOFS spectrometer at the CP-5 reactor at Argonne National Laboratory (Kleb et al. 1973). The TNTOFS instrument is a time-of-flight spectrometer which can be used either with a conventional Fermi chopper or with a statistical chopper. The statistical chopper has a duty cycle of ca. 50 % and is in general superior to the conventional chopper in applications where the total background on the detector is larger than twice the total signal that is obtained when the conventional chopper is used (Sköld 1968 a, b; Price & Sköld 1970). In the present experimental situation the background is two orders of magnitude larger than the signal for the conventional chopper and the statistical chopper therefore offers a significant advantage in this case.

The large absorption in the sample enhances the relative importance of extraneous scattering in the sample region, such as scattering from the sample container. With a strongly absorbing sample the only practical configuration is to scatter neutrons off the sample surface with the detectors in reflexion geometry. This arrangement is illustrated schematically in figure 1 for a container of conventional design and for the particular container used in the present studies. With sample in the container, neutrons are scattered in a thin layer at the surface of the sample

† 1 barn (b) = 10^{-28} m².

(the $1/e$ distance for 4 Å neutrons in liquid ^{3}He is *ca.* 0.05 mm) and also in the front wall of the container. The data must be corrected for the scattering in the front wall and this component is determined in a separate experiment in which the scattering from the empty container is measured. However, with the container empty, neutrons also strike the back wall of the container. For a container of conventional design, neutrons scattered off the back wall can reach the detectors. With the wedge-shaped container used in the present studies, neutrons striking the back wall cannot reach the detectors and the scattering from the front wall is in this case properly determined from a separate experiment with the container empty.

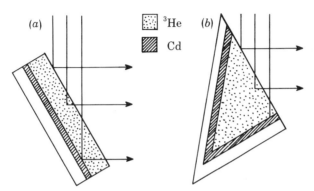

FIGURE 1. (a) Sample container of conventional design; (b) the particular design used in the present experiments.

The data must also be corrected for the energy dependent absorption in the sample. This correction is sensitive to the inclination of the sample surface to the incident and to the scattered neutron beams. As explained by Sköld *et al.* (1980), with a special device the orientation of the sample inside the cryostat is determined to within $\pm 0.25°$. To determine the scattering functions of an absolute scale, it is furthermore necessary to calibrate a number of spectrometer constants, such as the efficiencies of the beam monitors and of the detectors. This is accomplished by measuring the number of neutrons elastically scattered from a slab of vanadium into each detector for a given count on the beam monitors. The absolute normalization of the experimental scattering function is accurate to within *ca.* 10%.

The scattering functions are determined for $T = 40$ mK and for $T = 1.2$ K and in each case time-of-flight spectra are recorded at 15 angles of scattering in the range 13.2–$110.4°$. The incident neutron energy is 4.94 meV and the energy resolution of the instrument varies from 0.25 meV at small angles of scattering to 0.30 meV at large angles of scattering.

EXPERIMENTAL RESULTS

By using standard correction procedures (Copley *et al.* 1973), the experimentally observed time-of-flight spectra are converted to scattering functions at constant angle of scattering. As explained above, the observed functions contain contributions both from the density fluctuation scattering and from the spin fluctuation scattering, weighted by their respective bound scattering cross sections

$$S(q, E) = S^C(q, E) + (\sigma_i/\sigma_c) S^I(q, E), \qquad (11)$$

where $\sigma_c = 4.9$ b is assumed for the coherent cross section (Kitchens *et al.* 1974); the value for the spin-dependent scattering cross section, σ_i, is discussed below. Examples of experimentally determined scattering functions at constant angles of scattering are shown in figure 2 for the two temperatures of the experiment. In the constant-angle representation, the value of the wavevector transfer varies with the energy transfer, and selected values of q are shown at the top of the graphs in figure 2.

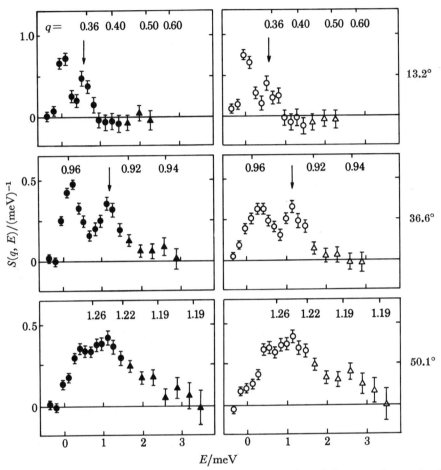

FIGURE 2. Selected examples of the scattering functions at constant value of the scattering angle. Data to the left are for $T = 40$ mK and data to the right are for $T = 1.2$ K. Vertical bars on the data points indicate the statistical uncertainty of the data points (± 1 standard deviation). Values of q are in reciprocal ångströms.

Analysis of the scattering functions in terms of theoretical models is best done at constant value of q, and such representations are presented and discussed below. It is useful, however, to identify important features in the results shown in figure 2, as these are more directly related to the actual observations. From the results shown in figure 2 the following general conclusions can be drawn. For small values of q the scattering functions show two peaks, one peak at *ca.* 0.1–0.2 meV, which is due to the excitation of particle–hole (p–h) pairs, and one peak at *ca.* 1 meV, which is the zero sound mode. For $q \gtrsim 1.3$ Å^{-1}, the two peaks merge, the zero sound mode approaches the upper edge of the p–h band and is Landau damped. The spectral weight of the tail at large energies, which is identified as multiple p–h excitations, increases with increasing q as expected. From the results shown in figure 2 it is clear that the zero sound mode continues to be a well defined excitation at 1.2 K while the p–h peak shows considerable broadening at the higher temperature. The thermal broadening is clearly displayed in the middle graphs in figure 2. For the data shown in the top graphs in figure 2 the shape of the spin fluctuation spectrum is dominated by the instrumental resolution and the broadening with temperature is therefore not obvious in this case ($\Delta E_{\text{res}} \approx 0.25$ meV).

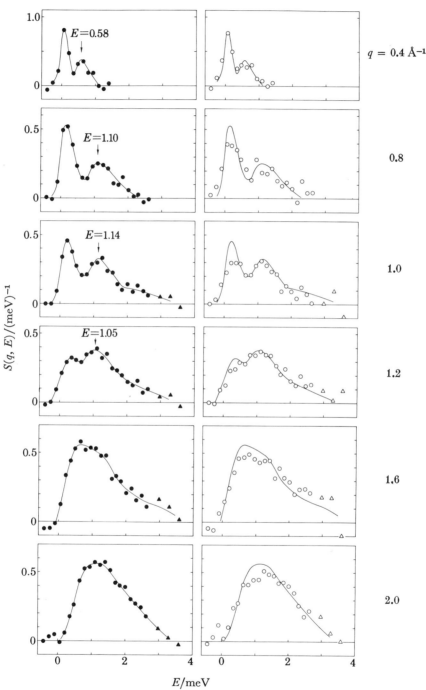

FIGURE 3. Selected examples of the scattering function at constant value of the scattering vector, q. The solid lines are fitted by eye to the results for $T = 40$ mK (left) and are also shown together with the results for 1.2 K (right).

To facilitate comparison with theoretical models, it is convenient to consider the scattering function at constant value of q rather than at constant angle of scattering. The constant q representation is derived by interpolation of the functions measured at constant angles of scattering. For each energy transfer, a cubic spline is fitted by least squares to the 15 constant-angle data points and the spline function is then used as a representation of the scattering function at arbitrary values of q. For a detailed account of the interpolation procedure, see Sköld et al. (1980). Examples of the constant q results are shown in figure 3. The solid curves in figure 3 are fitted by eye to the 40 mK data and are also shown with the 1.2 K data to demonstrate the change in the scattering function with temperature. The main features of the constant q curves are similar to those already observed in the constant-angle data, namely, for $q \lesssim 1.3$ Å^{-1} the scattering function shows two peaks plus a tail at large energies. The major difference between the data at 40 mK and at 1.2 K is the broadening observed in the particle–hole spectrum. The interpretation of the results in quantitative terms is discussed in the next section.

Theoretical interpretation

For detailed analysis of the experimental results we concentrate on the data for $T = 40$ mK. As noted above, the major difference between these results and those observed at 1.2 K is the broadening of the particle–hole spectrum at the higher temperature. For a discussion of finite temperature effects, see recent publications by Glyde & Khanna (1977, 1980) and Aldrich & Pines (1978).

To extract quantitative information from the experimental results, a phenomenological function has been fitted by least squares to the observed scattering function. The phenomenological function is the sum of three components, namely one function that describes the p–h spectrum, one that describes the zero sound peak and one that accounts for the multi-pair (m.p.) spectrum at large energies.

From a consideration of the Landau limit of the R.P.A. theory, it is easy to show that the spectral weight of the density fluctuation spectrum inside the p–h band is negligible at small values of q (Aldrich et al. 1976a). The p–h spectrum is therefore dominated by spin fluctuation scattering and in the fitting procedure it is assumed that this is the case at all values of q considered ($q \leqslant 1.3$ Å^{-1}). This component is represented by the paramagnon model (equation (10)) with $\bar{I}$ and the weight as fitting parameters. The theoretical function is folded with the resolution function before being compared to the experimental points. The zero sound peak is represented by a Gaussian function with position, width and area as parameters. The tail at large energies is fitted by a Gaussian centred at 2 meV, and with f.w.h.m. equal to 2 meV. As the statistical accuracy of the data at large energies is rather poor, the spectral weight of this component has been adjusted such that the f-sum rule is satisfied for the density fluctuation spectrum. The paramagnon model exhausts the f-sum rule for the spin spectrum and it is therefore assumed that there is no spin contribution to the m.p. spectrum.

The general conclusion is that the phenomenological model described above allows a very precise description of the experimental results; a typical example of the agreement achieved is shown in figure 4. However, the result shown in figure 4 indicates that the three components to the scattering function overlap to a substantial degree. This implies that the numerical results obtained for the individual components are strongly correlated. To assess the uniqueness of the results obtained from the model-fitting procedure, another model for the spin fluctuation spectrum is therefore also considered. In this case the p–h band is fitted by the function appropriate to the non-interacting Fermi gas with $m_q^* = 3 m_0$ and the spectral weight as the only fitting parameter. The results of this model are shown together with those obtained from the paramagnon model whenever appropriate. The results obtained from this analysis are summarized in figure 5. The energy of the zero sound mode is shown in figure 5a, together with the p–h band. The disappearance of the zero sound mode at $q \approx 1.3$ Å^{-1} is seen to be consistent with $m_q^* = 3 m_0$. Landau damping (decay of the mode into p–h excitations) will act to damp the

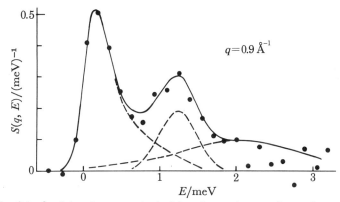

FIGURE 4. Example of the fit of the phenomenological function to the experimental scattering function as explained in the text. The solid line shows the sum of the component functions (broken curves).

mode very efficiently as the mode approaches the band edge and the position of the band edge is determined by m_q^*. The theoretical curve shown in figure 5a is the result obtained by Aldrich & Pines (1978) and is adjusted to fit the curve shown by the open circles, which were the only results available at that time. Although the present analysis of the experimental results suggests a need to re-evaluate the parameters in the R.P.A. model it is of interest to discuss the implications of the theoretical fit. The R.P.A. results are based on $r_c^{\uparrow\uparrow} = r_c^{\uparrow\downarrow} = 3.0$ Å, which should be compared with the value 2.68 Å which these authors obtain from a similar analysis of the excitation spectrum of liquid ^{4}He (Aldrich & Pines 1976; Aldrich et al. 1976b). They speculated that the difference between liquid ^{3}He and liquid ^{4}He is due to more effective short-range screening of the bare interaction in the case of liquid ^{3}He, which has a much larger zero point amplitude. Aldrich & Pines (1978) suggest that an experiment on liquid ^{3}He at elevated pressure, such that the zero point amplitude is reduced to a value comparable to that in liquid ^{4}He at s.v.p., may help to verify this hypothesis. Such experiments have been attempted (Hilton et al. 1978; Sköld & Pelizzari 1978b) but so far results that are accurate enough to be useful in this regard have not been obtained. The pressure dependence of the zero sound mode has also been discussed recently by Glyde & Khanna (1980).

Figure 5b shows the spectral weight of the zero sound mode. As in the case of the mode energy, these results are also rather sensitive to the assumed shape of the p–h spectrum. However, the overall shape of the curves are rather similar in the two cases and in general agreement with calculations (Aldrich 1974). The sharp rise in the curve at ca. 1.2 Å^{-1} signifies the strong coupling to p–h excitations as the mode approaches the band edge.

The results for the width of the mode, shown in figure 5c, are perhaps more surprising. The increase for $q \gtrsim 1.2$ Å^{-1} is expected and is due to the onset of Landau damping. However, the local increase in the width for $q \approx 0.6$ Å^{-1} is a novel feature, not predicted by any of the theoretical models currently available. We believe that this phenomenon can be explained if reference is made to the dispersion curve shown in figure 5a, namely, the zero sound mode shows large positive dispersion in the range $0.5 \text{ Å}^{-1} \lesssim q \lesssim 0.8 \text{ Å}^{-1}$ and decay of the mode into two excitations of lower q, and E is then kinematically allowed. This process has already been considered for liquid ^{4}He where the positive dispersion is much less pronounced than that observed here. However, in view of the large uncertainties in the present results and in the absence of detailed calculations it is not meaningful to discuss this hypothesis quantitatively; it is therefore left as a qualitative conjecture at this time. The width of the zero sound mode at other values of q could be explained by damping through decay into multipair excitations. This process is not included in the R.P.A. calculations discussed above, as these calculations use screened response functions corresponding to single excitations only. Decay of the zero sound mode due to multipair interaction has, however, been considered by Glyde & Khanna (1980). Using the theory for sound propagation in a dilute Fermi gas that includes scattering of quasi-particles, they calculate that the width of the collective mode is 0.22 meV at $q = 0.5 \text{ Å}^{-1}$ and 0.35 meV at $q = 1.0 \text{ Å}^{-1}$. Assuming, as above, that the peak in the width curve at $q \approx 0.6 \text{ Å}^{-1}$ is due to multiphonon decay and that the increase in the width for $q \gtrsim 1.2 \text{ Å}^{-1}$ is due to Landau damping, the results obtained by Glyde & Khanna for the multipair damping are entirely consistent with the results shown by the solid circles in figure 5c. Note that the experimental results shown in figure 5c are corrected for resolution broadening and thus represent the true line width of the mode. It should also be noted that the overlap of the p–h band with the zero sound mode, such as shown in figure 4, does not suggest a damping mechanism as, for reasons of symmetry, single particle spin fluctuations do not couple to the density mode.

Figure 5d shows the area of the m.p. spectrum. With the assumptions made above, the total fitted function fulfils the f-sum rule. It is gratifying to note that the function is in very good agreement with the data (compare figure 4). This implies that the absolute normalization of the data is rather accurate. For small values of q the spectral weight of the m.p. spectrum should vary as q^4. The results shown in figure 5d are approximately consistent with this behaviour at all values of q.

The result for the spin fluctuation spectrum are shown in figure 5e, which shows the values obtained for the paramagnon parameter, $\bar{I}$, and for the weight function, σ_i/σ_c. For consistency the value of σ_i/σ_c should be the same at all values of q. The present results yield $\sigma_i/\sigma_c = 0.2 \pm 0.05$ for the average value. This is consistent with the result obtained previously from a direct determination of the spin structure factor, namely $\sigma_i/\sigma_c = 0.25$ (Sköld et al. 1976). The value for $\bar{I}$ is in the range 0.9–1.0, which should be compared with $\bar{I} = 0.895$, the value obtained from the enhancement in the static susceptibility, and to $I \approx 0.9$, the value obtained by Beal-Monod (1979) from an analysis of earlier neutron scattering data by the present authors. Within the accuracy of the methods used to extract I from the experimental scattering function, these results are all consistent and support the view that normal liquid ^{3}He is nearly ferromagnetic. It would be of considerable interest to extend the neutron scattering measurements to lower values of q and to improve the energy resolution such that the intrinsic shape of the spin fluctuation spectrum is resolved. This would allow a more definite comparison of the neutron scattering results to the predictions of the paramagnon model. It would also be of considerable interest to study the spin fluctuation spectrum at elevated pressure and to search for indications of a magnetic instability in the nuclear spin system at finite q, as discussed by Aldrich & Pines (1978).

ELEMENTARY EXCITATIONS IN LIQUID ³He

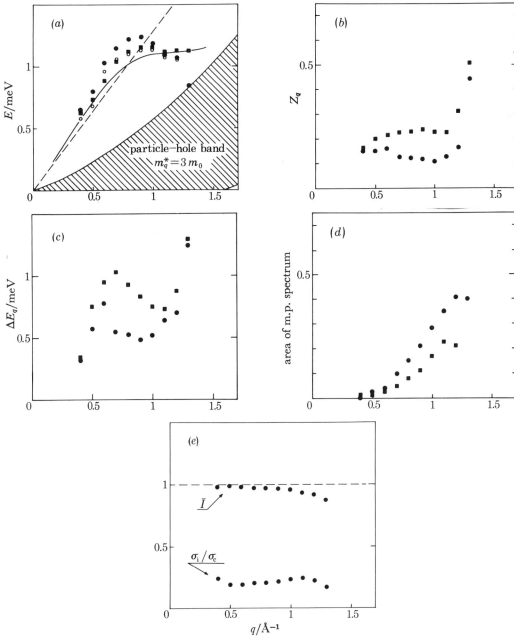

FIGURE 5. Parameters of the phenomenological function fitted to the experimental neutron scattering function for $T = 40$ mK. (a) Energy of the zero sound mode if the spin fluctuation peak is fitted by the paramagnon model (●) and if the spin fluctuation peak is fitted by the free Fermi gas model (■). Open circles show the results obtained if the peak positions are estimated by eye. (b) Spectral weight of the zero sound mode. Symbols as above. (c) Full width at half maximum of the zero sound mode. Symbols as above. (d) Area of the multi-pair spectrum. Symbols as above. (e) Parameters of the paramagnon model.

Discussion

The results discussed in this paper demonstrate that the neutron scattering results now available for liquid ³He are accurate enough to allow very detailed comparison to theoretical models. From a phenomenological decomposition of the scattering function, useful information about both the spin fluctuation spectrum and the density fluctuation spectrum is obtained.

The width of the collective density mode shows a complex behaviour as a function of the wavevector transfer. The results suggest that decay into single pair and multipair excitations as well as phonon–phonon decay may be important damping processes. The spin fluctuation spectrum is analysed in terms of the paramagnon model and a value for the paramagnon parameter close to unity is deduced.

For the future, it is suggested that measurements at elevated pressure would be of considerable interest. In the density fluctuation spectrum a variation of the average density would elucidate the importance of the zero point amplitude for the local screening of the effective repulsive pair interaction. In the spin fluctuation spectrum a variation of the density may lead to an enhancement in the spin-asymmetric interaction and, possibly, to a magnetic instability at finite q. It would also be of considerable interest to study the spin fluctuation spectrum with improved energy resolution such that the intrinsic shape of the paramagnon peak at small q is resolved. This would allow a detailed examination of the spin contribution to the effective pair interaction.

References (Sköld & Pelizzari)

Aldrich III, C. H. 1974 Ph.D. thesis, University of Illinois.
Aldrich III, C. H. & Pines, D. 1976 *J. low Temp. Phys.* **25**, 677–690.
Aldrich III, C. H. & Pines, D. 1978 *J. low Temp. Phys.* **32**, 698–715.
Aldrich III, C. H., Pethick, C. J. & Pines, D. 1976a *Phys. Rev. Lett.* **37**, 845–848.
Aldrich III, C. H., Pethick, C. J. & Pines, D. 1976b *J. low Temp. Phys.* **25**, 691–697.
Beal-Monod, M. T. 1979 *J. low Temp. Phys.* **37**, 123–134.
Copley, J. R. D., Price, D. L. & Rowe, J. M. 1973 *Nucl. Instrum. Meth.* **107**, 501–507.
Doniach, S. & Engelsberg, S. 1966 *Phys. Rev. Lett.* **17**, 750–753.
Glyde, H. R. & Khanna, F. C. 1977 *Can. J. Phys.* **55**, 1906–1923.
Glyde, H. R. & Khanna, F. C. 1980 *Can. J. Phys.* **58**, 343–350.
Hilton, P. A., Cowley, R. A., Stirling, W. G. & Scherm, R. 1978 *Z. Phys.* B **30**, 107–110.
Kitchens, T. A., Oversluizen, T., Passell, L. & Schermer, R. I. 1974 *Phys. Rev. Lett.* **32**, 791–794.
Kleb, R., Ostrowski, G. E., Price, D. L. & Rowe, J. M. 1973 *Nucl. Instrum. Meth.* **106**, 221–229.
Pines, D. 1966 In *Quantum fluids*, pp. 257–266. Amsterdam: North-Holland.
Price, D. & Sköld, K. 1970 *Nucl. Instrum. Meth.* **82**, 208–222.
Price, D. L. 1978 In *The physics of liquid and solid helium*, part 2, pp. 675–726. New York: John Wiley & Sons.
Scherm, R., Stirling, W. G., Woods, A. D. B., Cowley, R. A. & Coombs, G. I. 1974 *J. Phys.* C **7**, L341–345.
Sköld, K. 1968a *Bull. Am. phys. Soc.* (II), **13**, 467.
Sköld, K. 1968b *Nucl. Instrum. Meth.* **63**, 114–116.
Sköld, K. & Pelizzari, C. A. 1977 In *Quantum fluids and solids*, pp. 195–205. New York: Plenum.
Sköld, K. & Pelizzari, C. A. 1978a *J. Phys.* C **11**, L589–592.
Sköld, K. & Pelizzari, C. A. 1978b (Unpublished.)
Sköld, K., Pelizzari, C. A., Kleb, R. & Ostrowski, G. E. 1976 *Phys. Rev. Lett.* **37**, 842–845.
Sköld, K., Pelizzari, C. A., Kleb, R. & Ostrowski, G. E. 1980 (In preparation.)
Stirling, W. G., Scherm, R., Hilton, P. A. & Cowley, R. A. 1976 *J. Phys.* C **9**, 1643–1663.
Stirling, W. G., Scherm, R., Volino, F. & Cowley, R. A. 1975 In *Proceedings of the Fourteenth International Conference on Low Temperature Physics*, Otaniemi, Finland, pp. 76–79. Amsterdam: North-Holland.

The electric and magnetic moments of the neutron

By J. M. Pendlebury and K. Smith

*School of Mathematical and Physical Sciences, University of Sussex,
Falmer, Brighton, Sussex, BN1 9QH*

It is well known that the free neutron decays spontaneously into a proton, an electron and an antineutrino, that it has a spin of $\frac{1}{2}\hbar$ and a negative magnetic moment, but very careful measurements have failed as yet to reveal any evidence for a finite electric charge or dipole moment. This paper contains a brief discussion of early work and more detail of recent experiments at the Institut Laue–Langevin (I.L.L.), Grenoble, which have shown that the neutron charge is probably less than 4×10^{-20} electron charges (Bayreuth–Munich group), the neutron electric dipole moment (e.d.m.) is less than 1.5×10^{-24} cm times the electron charge (Oak Ridge–Harvard–Sussex group), and the ratio of the neutron and proton magnetic moments is equal to $-0.68497947(17)$, the uncertainty being only $0.25/10^6$ (Harvard–Sussex–Oak Ridge group). The main features of the Leningrad experiments with ultra-cold neutrons, which have reduced the neutron electric dipole length to 7.5×10^{-25} cm, are reported, with some details of the performance of the ultra-cold neutron magnetic resonance spectrometer now working at I.L.L. and the way it will be used to look for a neutron e.d.m. The paper concludes with some comments on the importance of the neutron moments to the development of the theory of fundamental particles.

Measurement of neutron charge

The earliest estimate of the neutron charge was made by Dee (1932), who studied ionization in gases due to neutrons and concluded that $q_n < 1.4 \times 10^{-3} q_e$. Twenty years later, Shapiro & Estulin (1957) looked for the sideways deflexion of a thermal neutron beam in an electric field and estimated that $q_n < 6 \times 10^{-12} q_e$. A big increase in sensitivity, by a factor of more than 10^6, was obtained by Shull *et al.* (1967) when they measured the effect of an electric field on the direction of the neutron beam passing between silicon crystals. Although they could detect deflexions as small as 5×10^{-6} rad, there was no significant correlation between electric fields as high as 225 kV cm^{-1} and the transmitted neutron intensity. They concluded that $q_n = -(1.9 \pm 3.7) \times 10^{-18} q_e$.

A still more sensitive experiment is now being carried out by R. Gähler, J. Kalus & W. Mampe (personal communication) at I.L.L., Grenoble, using the apparatus shown in figure 1. Partly monochromatized neutrons with wavelength 20.6 ± 0.5 Å† that pass the 50 μm entrance slit are focused on to the 30 μm exit slit 10 m away by the quartz lens. The optical bench with a mass of 4 t is water-filled for thermal stabilization, and the flight path, over which an electric field of 200 kV cm^{-1} can be applied, is evacuated to 10^{-4} Torr‡. Results obtained so far indicate that $q_n < 4 \times 10^{-20} q_e$ and that the limit is likely to be reduced by another factor of four by the time the calibration experiments have been completed.

Several methods of measuring the neutrality of molecules have been reviewed by Dylla &

† 1 Å = 10^{-10} m = 10^{-1} nm.   ‡ 1 Torr ≈ 133.3 Pa.

King (1973). The lowest limit has been reported by Hillas & Cranshaw (1960), who looked for a change in electric potential when a large mass of gas was allowed to leave a metal container and concluded that the charge on a helium atom is less than 5×10^{-21} electron charges.

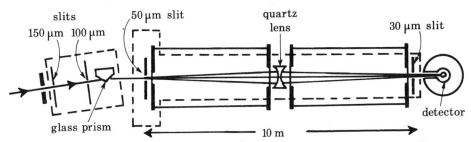

FIGURE 1. Schematic arrangement of the system used by the Bayreuth–Munich group at I.L.L., Grenoble, to obtain an upper limit for the neutron charge.

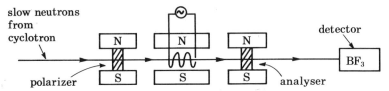

FIGURE 2. Apparatus used by Alvarez & Bloch to make the first measurement of the neutron magnetic moment in 1940.

MEASUREMENT OF THE NEUTRON MAGNETIC DIPOLE MOMENT

The first measurement of the neutron magnetic dipole moment μ_n was carried out by Alvarez & Bloch (1940) using the apparatus shown schematically in figure 2. Neutrons produced by D bombardment of Be in a cyclotron were slowed down and polarized by transmission through magnetized iron, allowed to pass through a few centimetres of steady transverse magnetic field with a perpendicular oscillating field, and then analysed by transmission through a second block of magnetized iron. The oscillating field induced a spin-flip with maximum probability when the rotating component had the same angular frequency and direction as the spin precession in the steady field and the neutron intensity at the detector fell by a small percentage when the resonant condition $h\nu = 2\mu_n B$ was satisfied. The resonance curve obtained had a width roughly equal to $1/T$, where T was the average time spent by the neutrons in the oscillating field. The magnitude of the magnetic moment obtained was $1.93 \pm 0.02\,\mu_N$, the accuracy being limited mainly by the difficulty of measuring the steady magnetic field by pick-up coil techniques. No indication of the sign of the moment was obtained because an oscillating rather than a rotating field was used.

A considerable increase in accuracy was obtained using similar apparatus by Arnold & Roberts in 1947 and, independently, by Bloch et al. in 1948, by using nuclear magnetic resonance techniques to measure the steady magnetic field in terms of the proton precession frequency, thereby allowing a measurement of the ratio of the neutron and proton precession frequencies in a given field, and hence the ratio of the magnetic moments, without introducing the errors associated with absolute field measurements. Bloch et al. obtained the result $|\mu_n/\mu_p| = 0.685001(30)$. In 1949, Rogers & Staub used a rotating magnetic field in place of the oscillating field, obtained the neutron resonance with only one sense of rotation, and showed that the neutron magnetic moment is negative.

An important technical advance was introduced by Ramsey (1949) who showed that the precision of a beam resonance experiment can be improved by using short oscillating field regions at the beginning and end of the steady field section instead of a single oscillating field over the whole region. The first oscillating field, as shown in figure 3, rotates the spin into the plane perpendicular to the steady field, a $\frac{1}{2}\pi$ spin flip, the spin then precesess about the steady field until the second oscillating field induces a second $\frac{1}{2}\pi$ spin-flip. The overall probability of spin-flip by π is then a damped oscillating function of the oscillating field frequency with a central oscillation width of $1/2T$, where T is the transit time between the oscillating field regions, and an overall envelope of width $1/\tau$, where τ is the duration of each oscillating field. The centre of the pattern comes at a frequency equal to the average of the precession frequency between the oscillating field regions, and the spin-flip probability is unity on resonance when the oscillating field has the optimum value.

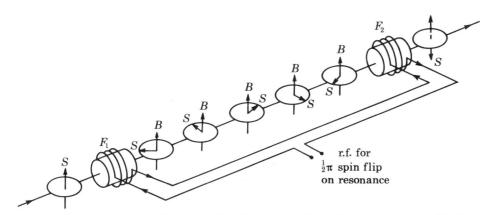

FIGURE 3. The Ramsey separated oscillatory field technique, which results in a resonance with the centre of the pattern at a frequency equal to the average over the transit time between the two oscillating fields.

The double field technique was applied by Ramsey and coworkers in 1954 to the measurement of the neutron magnetic moment with the use of the apparatus shown in figure 4. Reflexion from magnetized iron at glancing incidence was used to polarize the neutrons, and they obtained the resonance curve shown in figure 5 when they plotted against frequency the intensity change when the phase of the second oscillating field was reversed. The magnetic moment ratio obtained after correcting for magnetic shielding of the protons was $|\mu_n/\mu_p| = 0.685039(17)$ with an accuracy of $25/10^6$, which is still limited mainly by the difficulty of averaging effectively the magnetic field between the oscillating field regions.

An experiment recently carried out by Greene et al. (1977) at I.L.L., Grenoble, has largely overcome this problem and resulted in a 100-fold improvement in accuracy. A flowing water technique allows the precessing protons to sample almost the same magnetic field as the neutrons and yield a proton resonance curve with the centre at the average of the proton precession frequency over the region between the two oscillating fields. The system used is shown schematically in figure 6. The flowing water, which has a longitudinal relaxation time of several seconds, is polarized nearly to the equilibrium value by taking about 6 s to pass through a baffled container in a field of 2000 G†, it then flows through the oscillating and steady field regions, where the field may well be only a few gauss, into an n.m.r. coil in a field of 2600 G,

† $1\text{ G} = 10^{-4}\text{ T}$.

where the remaining proton polarization is sampled by using a detector tuned to the proton resonance in the strong field. The variation of the n.m.r. detector reading with the frequency of the oscillating field then shows the typical oscillatory shape of the Ramsey system with the frequency of the centre of the pattern at the average of the proton frequency between the two

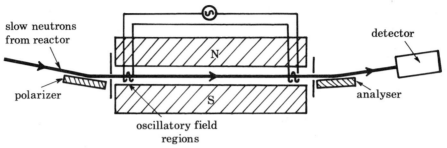

FIGURE 4. Neutron magnetic resonance apparatus with separated oscillatory fields and magnetized iron–cobalt mirrors as polarizer and analyser.

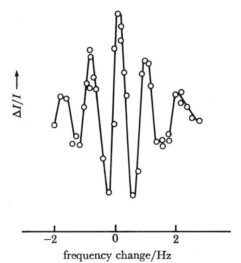

FIGURE 5. Neutron magnetic resonance obtained by Corngold, Cohen and Ramsey using separated oscillatory fields.

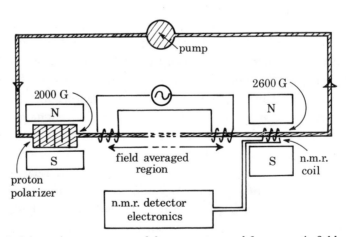

FIGURE 6. Schematic arrangement of the apparatus used for magnetic field averaging by the flowing water technique.

oscillating field regions. Greene used a neutron system similar to that of figure 4 with the exception that the neutrons passed from the polarizer through the oscillating field regions to the analyser within a 1 cm diameter glass tube acting as a neutron guide, thereby avoiding intensity loss because of beam divergence, and the pipes normally carrying water passed through the same oscillating field coils above and below the neutron guide. Typical proton

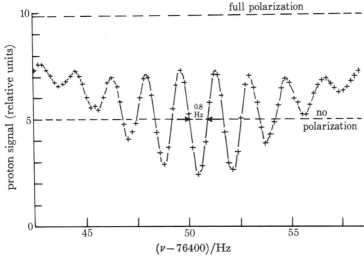

FIGURE 7. Proton resonance signal obtained by the flowing water technique. The line width is approximately $1/2T$ where T is the transit time between the two oscillating fields.

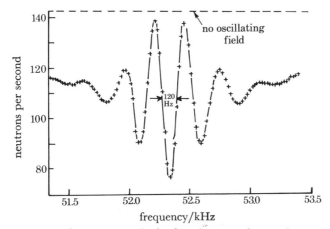

FIGURE 8. Neutron magnetic resonance obtained at the same time as the proton resonance in figure 7 by connecting the oscillating field coils to both signal generators.

and neutron resonances are shown in figures 7 and 8 respectively, the line width in each case being roughly $1/2T$ where T is the transit time between the oscillating field coils, the very narrow proton line arising from the slow flow rate of the water. After checking the system by sending water instead of neutrons down the centre tube and taking into account effects due to the Bloch–Siegert effect, oscillatory field phase errors, oscillating field inhomogeneity, neutron velocity distribution and the change in the monitoring tube field by water in the guide tube, the magnetic moment ratio obtained was $|\mu_n/\mu_p| = 0.68497947(17)$, with an error of $0.25/10^6$.

Measurement of the neutron electric dipole moment

The Hamiltonian for a neutron with magnetic moment μ_n and electric dipole moment p_n in a magnetic field B and an electric field E is

$$\mathcal{H} = -\mu_n \cdot B - p_n \cdot E,$$

and the frequency, ν, at which resonance would occur in a neutron resonance apparatus such as that shown in figure 4 with parallel E and B fields is given by

$$h\nu = -2\mu_n B - 2p_n E.$$

The change in resonance frequency that results when the direction of the E field is reversed relative to the B field is then independent of the B field, and one obtains

$$h\Delta\nu = -4p_n E = -4q_e DE,$$

where q_e is the electron charge and D is the electric dipole length, thereby allowing a direct measurement of the dipole length in terms of the electric field applied and the frequency shift.

The first attempt to measure the neutron e.d.m. in this way was made by Smith *et al.* (1951), who showed that $D < 5 \times 10^{-20}$ cm, an e.d.m. that would produce a frequency shift of only 10 Hz when the electric field was 200 kV cm^{-1}, a small proportion of the resonance curve line width, which would have been about 1000 Hz. The experiment was therefore carried out by adjusting the steady magnetic field and the frequency of the oscillating field to give a working point on the steepest part of the resonance curve and then looking for correlation between the neutron count rate and the direction of the electric field relative to the magnetic field.

Similar experiments have been carried out since 1951 by several groups in America and Europe, the most recent being that completed at I.L.L., Grenoble, by Dress *et al.* (1977). Using the neutron magnetic resonance apparatus which was subsequently modified for the precision μ_n/μ_p work already mentioned, they obtained neutron resonances similar to figure 8 with a width of 45 Hz. A magnetic shield helped to reduce the sensitivity of the system to external magnetic fields, and a computer was used to run the experiment and reduce to a minimum the effects of magnetic field drifts, reactor power drifts and changes in alignment of the 3 m long apparatus. The most important systematic problem was the effective magnetic field $E \wedge v/c$ seen by neutrons moving with velocity v through the electric field E, a magnetic field which produced frequency shifts and associated count rate changes correlated with the direction of the electric field and therefore difficult to distinguish from a real e.d.m. effect. To compensate reliably for this effect, the whole apparatus, including the mirrors and shield, was rotated about a central axis, leaving behind only the reactor and the detector, thereby producing an effective reversal of v. The final result of $(0.4 \pm 1.5) \times 10^{-24}$ cm for the dipole length probably represents the limit of the beam experiments, for the electric field cannot be increased significantly above 200 kV cm^{-1}, and the apparatus cannot be made much longer than 3 m so it is unlikely that the line width can be reduced below about 10 Hz. Furthermore, the $E \wedge v/c$ effect will become increasingly important as the limit is reduced, unless there is a corresponding reduction in v.

The experimental demonstration by Lushchikov *et al.* (1969) that neutrons with velocities less than 6.5 m s^{-1}, what are now called ultra-cold neutrons (u.c.n.), are reflected at normal incidence by many materials, has led to the possibility of storing polarized neutrons for many seconds and the production of neutron magnetic resonances with a line width $1/2T$, where T is the storage time, of less than 0.1 Hz. Furthermore, because of the small distance between the input and output ports in the storage vessel and the long storage time, the effective velocity in the $\boldsymbol{E} \wedge \boldsymbol{v}/c$ effect is very small and may be neglected.

A first attempt to measure the neutron e.d.m. by using u.c.n. has been made by Altarev et al. (1978) at Leningrad with the apparatus shown schematically in figure 9. The u.c.n. from the reactor are polarized by transmission through the magnetized iron foil; they then pass through

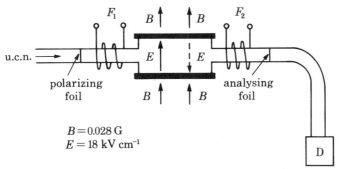

FIGURE 9. Arrangement of apparatus used by the Leningrad group to measure the neutron e.d.m. with stored ultra-cold neutrons. The input and exit apertures in the actual apparatus were at 90° to each other and the u.c.n. took on average about 6 s to bounce on the walls from one aperture to the other.

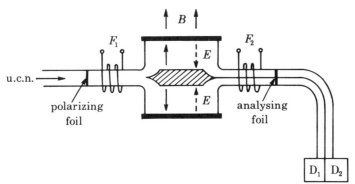

FIGURE 10. The double cavity system used by the Leningrad group to minimize the effects of magnetic field drifts.

the $\tfrac{1}{2}\pi$ spin-flipper into the storage volume where the magnetic field is 0.028 G and the electric field is 18 kV cm^{-1}. After a storage time, which averages about 6 s, the time needed for incoming neutrons to make collisions with the walls and eventually find the exit port, which is at 90° to the input port in the actual apparatus, the u.c.n. pass through the second $\tfrac{1}{2}\pi$ spin-flipper and the analysing foil on the way to the detector. The line width obtained is about 0.08 Hz, very much narrower than the resonances obtained in the beam experiments, but the gain is largely offset by the reduction in E by a factor of 10, the very large reduction in neutron count rate and the need to run for long periods to reach an e.d.m. limit due to systematic effects rather than counting statistics. To minimize the effects of external magnetic field changes, which become increasingly difficult to cancel as the line width is reduced, a double cavity spectrometer arranged like the system in figure 10 was used by the Leningrad group with E fields in opposite directions in the two cavities. Significant cancelling of the effect of field changes was obtained by using the difference between the two count rates when looking for correlations with reversals of the electric field. The dipole length reported is $D = (4.0 \pm 7.5) \times 10^{-25}$ cm, the accuracy being limited mainly by the changing magnetic fields owing to leakage currents in the insulating walls of the storage vessel.

A second u.c.n. magnetic resonance spectrometer is now under construction at I.L.L., Grenoble, with the layout shown schematically in figure 11. Polarized u.c.n. entering the

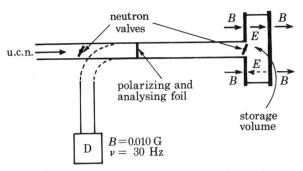

FIGURE 11. The system being used by the Sussex–Harvard–Rutherford group at I.L.L., Grenoble, to study neutron magnetic resonance with stored neutrons.

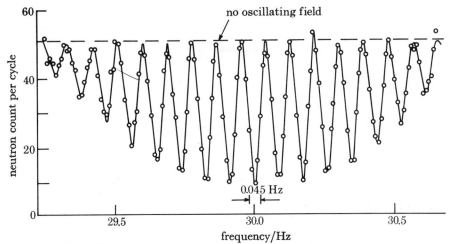

FIGURE 12. Neutron magnetic resonance obtained by applying two 1 s bursts of oscillating field separated by a 10 s period without any oscillating field.

storage volume through the neutron valve are stored with the valve closed for times that we hope will be as long as 100 s. Pulses of oscillating field applied perpendicular to the steady field at the beginning and end of the storage period produce the necessary $\tfrac{1}{2}\pi$ spin flips and the neutrons that leave when the valve is opened at the end of the storage period go through the analysing foil on the way to the detector. A resonance obtained in a magnetic field of 0.01 G with a storage time of 10 s is shown in figure 12. It is hoped that the problem of magnetic field drift and to some extent the leakage current effect can be minimized by monitoring the magnetic field in the cavity while the neutrons are being stored. Professor Ramsey has proposed to do this by storing polarized ^{3}He atoms in the storage volume with the neutrons and by using an n.m.r. technique to analyse the ³He atoms when they emerge with the neutrons at the end of the storage period. It is also hoped that it will be possible to improve the u.c.n. density in the e.d.m. apparatus by a factor of 100 or more by using as the source the u.c.n. that remain and build up in a liquid helium filled storage vessel at 0.8 K after down-scattering from an intense cold neutron beam, the development of a suggestion made by Golub & Pendlebury (1975, 1979).

THE NEUTRON MOMENTS AND NUCLEAR THEORY

If we assume that the fundamental particles introduced to explain the behaviour of matter are always identical particles, that total electric charge is always conserved in any observable process and that the electric charges of the fundamental particles remain the same when they are in combination, we conclude that the photon must have zero charge and that the charges on the electron and positron must be equal and opposite. We also conclude that the neutron charge must be equal to the sum of the electron, proton and antineutrino charges. Since the atomic charge can be written

$$q(Z, A) = Z(q_p + q_e) + (A - Z) q_n,$$

the combination of data for atoms with different $Z/(A-Z)$ ratios allows independent estimates of $q_p + q_e$ and q_n. Akhiezer & Rekalo (1975) conclude in their review on the electric charges of fundamental particles that the atomic neutrality data show that $(q_n + q_p) = (-8 \pm 5) \times 10^{-20} q_e$ and $q_n = (7 \pm 6) \times 10^{-20} q_e$. The experimental observation that q_n is less than $4 \times 10^{-20} q_e$ implies an experimental lower limit on the antineutrino charge of about $10^{-19} q_e$, the same as the proton–electron charge difference deduced from atomic neutrality measurements.

Application of the theory of quantum mechanics to fundamental particles shows that no fundamental particle can have an e.d.m. unless there are both simultaneous parity and time-reversal-violating interactions in particle physics. Parity violation in weak interactions such as those responsible for neutron decay has been well established but there is as yet no direct evidence for the violation of time reversal symmetry. The observation of a neutron e.d.m. would provide such evidence and also give important information about the type of interaction. Many calculations of the neutron e.d.m. have been carried out in the last 10 years assuming various types of P and T invariance, predicting neutron dipole lengths from 10^{-20} cm for electromagnetic interactions down to less than 10^{-32} cm for superweak interactions. The results of the theories are reviewed by Golub & Pendlebury (1972), a paper that contains an elementary treatment of P and T reversal, and Dress et al. (1977) discuss the recent neutron beam e.d.m. measurement. Extension of the neutron dipole length down to 10^{-26} cm would rule out the proposed electromagnetic and $\Delta S = 0$ weak interactions, and would aid significantly the refinement of existing theories.

Although there is no possibility of predicting the neutron magnetic moment with an accuracy of $0.25/10^6$ from nuclear theory, the existence of the magnetic moment implies circulation of charge within the neutron. Early measurements of the magnetic moments of light nuclei with small percentage accuracy showed that the neutron appears to have an effective magnetic moment of about $-1.91 \mu_N$ when it forms part of a nuclear system, and a simple single-particle model of the deuteron was able to account for the experimental observation that the deuteron moment is not equal to the sum of the neutron and proton moments. The only significant development in the theory of nucleon moments since 1950 is the proposal by Beg *et al.* (1964), and independently by Sakika (1964), that if quarks have spins and the internal symmetry group SU(3) of the baryons is broken only by electromagnetism, the neutron–proton magnetic moment ratio should be $-\frac{2}{3}$, a result close to but well outside the error in the experimental result $-0.68497945(17)$. The recent experiment at I.L.L. was not carried out to test nuclear theory! The e.d.m. apparatus already existed and could be easily modified, and the experiment was able to demonstrate the effectiveness of the flowing water technique of magnetic field averaging.

References (Pendlebury & Smith)

Akhiezer, A. I. & Rekalo, M. P. 1975 *Usp. Fiz. Nauk* **114**, 487–508.
Altarev *et al.* 1978 Submitted to *Nucl. Phys.*
Alvarez, L. W. & Bloch, F. 1940 *Phys. Rev.* **57**, 111.
Arnold, W. R. & Roberts, A. 1947 *Phys. Rev.* **71**, 878.
Beg, M. A., Lee, B. W. & Pais, A. 1964 *Phys. Rev. Lett.* **13**, 514.
Bloch, F., Nicodemus, D. B. & Staub, H. H. 1948 *Phys. Rev.* **74**, 1025.
Dee, P. I. 1932 *Proc. R. Soc. Lond.* A **136**, 727.
Dylla, H. F. & King, J. G. 1973 *Phys. Rev.* A **7**, 1224–1229.
Dress, W. B., Miller, P. D., Pendlebury, J. M., Perrin, P. & Ramsey, N. F. 1977 *J. Phys.* D **15**, 9.
Golub, R. & Pendlebury, J. M. 1972 *Contemp. Phys.* **13**, 519–558.
Golub, R. & Pendlebury, J. M. 1975 *Phys. Lett.* A **53**, 133–135.
Golub, R. & Pendlebury, J. M. 1979 *Rep. Prog. Phys.* **42**, 439–501.
Greene, G. L., Ramsey, N. F., Mampe, W., Pendlebury, J. M., Smith, K., Dress, W. D., Miller, P. D. & Perrin, P. 1977 *Phys. Lett.* B **71**, 297–300.
Hillas, A. M. & Cranshaw 1960 *Nature, Lond.* **186**, 459.
Lushchikov, V. I., Pokotilovsky, Y. N., Strelkov, A. V. & Shapiro, F. L. 1969 *JETP Lett.* **9**, 23.
Ramsey, N. F. 1949 *Phys. Rev.* **76**, 996.
Rogers, E. H. & Staub, H. H. 1949 *Phys. Rev.* **76**, 480.
Sakita, B. 1964 *Phys. Rev. Lett.* **13**, 643.
Shapiro, I. S. & Estulin, I. V. 1957 *Soviet Phys. JETP* **3**, 626.
Shull, C. G., Billman, K. W. & Wedgwood, F. A. 1967 *Phys. Rev.* **5**, 1415–1422.
Smith, J. H., Purcell, E. M. & Ramsey, N. F. 1951 *Phys. Rev.* **108**, 120.

The use of neutrons to study protein–RNA interactions

By B. Jacrot

Institut Max von Laue – Paul Langevin,
156X Centre de Tri, 38042 Grenoble Cédex, France

Protein–RNA interactions play a key role in the structure, morphogenesis and function of various systems (viruses, ribosomes and, more generally, protein synthesis). The neutron is a powerful tool to study those interactions. Some examples, are given. For viruses, neutrons provide structural information on the two molecules where they interact. Viral proteins do not appear to be simple globular proteins. In the interactions between tRNA and aminoacyl tRNA synthetases, neutrons allow a simultaneous study of the reaction and of the structural modifications associated with that reaction, giving a hint on the role of both electrostatic and specific interactions.

Introduction

Among the interactions that play a key role in cell life, those between ribonucleic acids (RNA) and protein are specially important. RNA is found associated with many functions in gene expression in the cell, and in all cases interactions with specific proteins play a key role. This may be at several levels. First, it may be a structural role, as for instance in a ribosome, where one of the actions of the protein–RNA interaction is to establish the structure of the particle so that it will perform its functions correctly. The role of the interaction may be also directly at the functional level, as for instance in the control of the RNA messenger. Another example of this last case is that of the reactions of tRNA with various enzymes involved in protein biosynthesis, such as aminoacyl tRNA synthetase, elongation factor and formylase.

Nature of the interaction

RNA is a genuine polyelectrolyte and this is expected to play a key role in its interaction with protein. For instance, basic groups (lysine and arginine), which are positively charged, having strong electrostatic links with the negatively charged phosphates of the RNA. This type of interaction is highly salt-dependent and does not vary with the sequence of the RNA. Here we consider single-stranded RNA. Such RNA has a secondary structure formed by base pairing (cooperative hydrogen bonding) between different parts of the molecule. This structure is complex and strongly dependent upon the sequence. It has been suggested (Weidner *et al.* 1977) that a given sequence may have different secondary structures that have nearly the same energy. The RNA is expected to be folded by various forces (salt links, hydrophobic, etc.) into a three-dimensional structure, as already established for tRNA. This makes an important difference from double-stranded DNA, which has broadly speaking a conformation independent of its sequence, which modulates only the exact organization of the double helix. Single-stranded RNA should have a three-dimensional structure that depends strongly on its composition and sequence and which may be modified by external parameters. This gives the possibility of very specific and precise interactions between RNA and the proteins that are functionally associated with it.

So far very little is known of the exact nature of those interactions. The normal approach would be to crystallize a complex between an RNA and interacting protein (for instance a complex between a tRNA and the corresponding aminoacyl tRNA synthetase). So far this has not been possible; the only crystallized systems in which RNA and proteins are in contact are viruses, and for that reason various groups (Harrison et al. 1978; Unge et al. 1979; Suck et al. 1978) have undertaken crystallographic studies of viruses. In the most advanced work (Harrison et al. 1978) a structure of tomato bushy stunt virus at 2.9 Å resolution has been obtained. This study gives much important information on protein–protein interactions, but none on protein–RNA interaction, as no density, associated with RNA, is found in the electron density map, indicating that the RNA is completely disordered. Owing to this lack of precise crystallographic information, other methods must be sought. Neutron scattering turns out to be one of the most useful.

Neutron scattering applied to nucleoprotein systems

We shall consider here only neutron scattering from solutions. This method with special references to biological applications, has been described in some detail (Jacrot 1976), and only the main points will be summarized.

In dealing with scattering from objects in solution, since only low resolution structural information can be obtained, the relevant quantity is the scattering density, which is defined by

$$\rho(r) = \frac{1}{V} \sum_v b_i. \tag{1}$$

The summation is done on a volume V large compared with interatomic distances, but small compared with the resolution of the data; b_i are the scattering amplitudes of all atoms included in the volume v. Indeed, as we consider objects in solution, the quantity that will come into the equations is the difference between the radial density distribution $\rho(r)$ of the object and that, ρ_s, of the solvent. In all practical cases the solvent for biological molecules is water with small amount of salts, and can be considered, from the neutron point of view, as pure water with a negative scattering density of 0.562×10^{-14} cm/Å^3†. If heavy water is substituted for ordinary water the scattering density of the solvent will rise to 6.4×10^{-14} cm/Å^3. In comparison the scattering density of a protein is about 1.8×10^{-14} cm/Å^3 and that of a nucleic acid around 4×10^{-14} cm/Å^3. The density of a protein (or a nucleic acid) increases somewhat with the amount of D_2O in the solvent. This is due to the exchange of labile protons. The contribution to the scattering curve of the two components (protein and nucleic acid) will vary with the amount of D_2O in the solvent. If the protein has a uniform density (this is never strictly true) in a solvent with some 40–42 % D_2O, only the RNA will contribute to the scattering. The reverse will be true in a solvent with 68–70 % D_2O. This simple fact is the basis of the study of nucleoproteins with neutrons. More generally, one defines the contrast of a particle as the difference between its average scattering density and that of the solvent. This contrast may be positive or negative.

† $1 \text{ Å} = 10^{-10} \text{ m} = 10^{-1} \text{ nm}$.

The intensity at the origin

A dilute solution of N identical particles gives a scattering curve which extrapolates at zero angle to

$$I(0) = (\sum b - b_s V)^2 N. \qquad (2)$$

The summation is over all atoms in the particles, and V is the volume occupied by those atoms (volume from which the solvent is excluded), and is related to the specific volume $\bar{v}$ of the particle of molecular mass M by

$$V = \bar{N}^{-1} M \bar{v},$$

where $\bar{N}$ is the Avogadro number. This relation is valid only if the system is a two-phase system: molecule and solvent. If the solvent in contact with the molecule is different from the bulk solvent (2) must be modified to take this into account. If the particle is a protein, b and V are proportional to M, whereas for a concentration expressed in mass per volume N is inversely proportional to M. Thus $I(0)$ is proportional to the molecular mass of the protein. Intensities are easily calibrated by scattering from water, and so the protein molecular mass can be measured easily (Jacrot & Zaccaï 1980).

Now if one considers a reacting system made of proteins A and nucleic acid B with, for instance, the equilibrium

$$A + B \rightleftharpoons AB,$$

the solution will be composed of a mixture A, B and AB defined by the equilibrium constant of the reaction, and the intensity at the origin will be given by

$$I(0) = (\sum b_A - b_s V_A)^2 N_A + (\sum b_B - b_s V_B)^2 N_B + (\sum b_A + \sum b_B - b_s V_A - b_s V_B)^2 N_{AB}. \qquad (3)$$

So if one starts with a pure solution of protein A, by adding successive quantities of nucleic acid B to A it will be possible to determine the equilibrium constant of the reaction, and more important to establish its stoichiometry. As we shall see later, this turns out to be the most reliable method to establish this stoichiometry, and its dependence on parameters such as ionic strength. The reaction can be followed in any solvent, but the stoichiometry will be better followed in H_2O where both components will contribute. If, on the other hand, one wishes to follow the behaviour of the protein moiety alone, this will be better done in a solvent where the contribution of the nucleic acid is minimized.

The radius of gyration

The intensity at very small angles is given by the Guinier approximation:

$$I(s) = I(0) \exp\left(-\tfrac{4}{3}\pi^2 s^2 R_G^2\right), \qquad (4)$$

where s is the scattering vector of length $(2/\lambda) \sin \tfrac{1}{2}\theta$ where θ is the scattering angle and λ the neutron wavelength. This will give the radius of gyration, R_G, of the particle. For a particle with uniform scattering density, R_G will be independent of contrast, but for a nucleoprotein in which the density of the RNA is higher than that of the protein, R_G will vary, as shown by Stuhrmann (1974), and this variation will provide some information on the relative distribution of RNA and proteins within the particle. This has been applied to ribosomal subunits (see reviews by Koch & Stuhrmann (1979) and Serdyuk (1979)).

If one considers again the reaction between a protein and a nucleic acid, it will be possible, by using the appropriate solvent, to follow changes of structure in one of the components during a reaction, characterized quantitatively by a change of radius of gyration.

The scattering at larger angles

Beyond the domain of validity of Guinier approximation, the scattering curve $I(s)$ is given by

$$I(s) = \left\langle \left| \int_v (\rho(r) - \rho_s) \exp(2\pi i s r) \, d^3 r \right| \right\rangle, \tag{5}$$

where the bracket represents the averaging over all orientations of the particles, which in a dilute solution are always randomly orientated. It is obvious that this averaging makes impossible, in the general case, the inversion of (5) to get $\rho(r)$ and this equation can only be used to test models of structure. Here again, the measurement of $I(s)$ with various contrasts allows models of the structure of both the RNA and the protein in a nucleoprotein particle to be tested. This has been done for the 30 S subunit of ribosomes (Serdyuk et al. 1979). A model is characterized by a certain number of parameters. Information theory gives the maximum number of parameters that can be obtained from a set of curves $I(s)$ at various contrast (for the case of an infinitely accurate experiment; in practice, the limited accuracy of the data will always limit the number of parameters to a smaller value than given by theory). This point is dealt by Luzzati (1979), giving for this number of parameters

$$J = 6 + 6 s_{\max} D, \tag{6}$$

where D is the largest dimension of the object and $s_{\max}$ the largest scattering vector in experimental data.

For a spherical object, the averaging over orientations disappears and (5) can be inverted. This applies to isometric icosahedral viruses, which, at least up to resolution comparable with distances between subunits, can be considered as spherical. If data are collected up to $s = (50\,\text{Å})^{-1}$ on virus of diameter 250 Å, (6) says that a maximum of 36 parameters can be obtained. Even taking into account the reduction of that number from data inaccuracy this is enough to analyse a spherical virus into four or five spherical shells (12 or 15 parameters), giving a rather detailed view of the relative organization of the RNA and the protein shell inside the virus.

Tomato bushy stunt virus (TBSV)

The method outlined above has been used to get a low-resolution structure of the virus TBSV (Chauvin et al. 1978). The main result of that study was to reveal an unfolding of the polypeptide chain, or a strand about ten amino acids long, in that part of the virus where most of the RNA is concentrated. This is shown in figure 1. This observation suggests the following comments.

(a) The inner part of the viral protein, being non-globular, may have several slightly different conformations. This gives rise to an internal disorder, which explains why this inner part of the virus is not observed by X-ray high-resolution crystallography.

(b) If protein–RNA interactions in the virus were purely electrostatic, this disorder would be unnecessary; on the other hand, if there are specific recognitions between an amino acid (or an amino acid sequence) and a nucleotide (or a nucleotide sequence), such disorder is necessary. Without that disorder the same amino acids would be found in positions related by the icosahedral symmetry, whereas the lack of periodicity in the nucleotide sequence makes difficult a folding of the RNA strand which would also bring identical nucleotides into symmetry-related positions in the virion.

(c) When the virus swells (J. Witz & B. Jacrot, unpublished results) at high pH in the presence of EDTA, this internal organization of the virus is preserved, showing its importance for the virus's stability.

(d) In another virus (brome mosaic virus), we have strong evidence that the viral protein is not globular and has a somewhat similar internal organization (M. Cuillel, B. Jacrot & M. Zulauf, unpublished results), although the protein does not go so far into the interior of the virion.

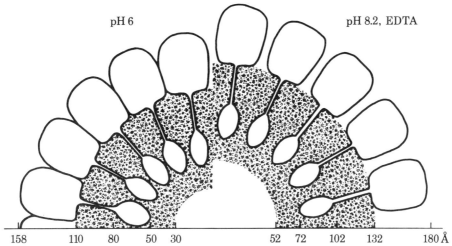

FIGURE 1. A schematic representation of the internal structure of tomato bushy stunt virus in its native state (left side) and in its swollen state (right side). The shadowed area represents regions occupied by RNA. The densities found in the swollen virus compared with those in the native one suggest that there is no reorganization during swelling but simply a displacement by 22 Å of a group of proteins (most likely trimers) with the RNA bound to it.

THE INTERACTION BETWEEN tRNA AND AMINOACYL tRNA SYNTHETASES

During protein synthesis, each tRNA must be charged with its appropriate amino acid. This is done by a group of enzymes (one for each tRNA) named aminoacyl tRNA synthetases. The reproducibility and precision of each protein synthesis is dependent on the proper recognition between each tRNA and the corresponding enzyme. This is a typical case where protein–RNA interaction plays a key role. Several systems have been investigated with neutron scattering (Dessen et al. 1978; Zaccaï et al. 1979; Zaccaï, private communication). Data were collected by titrating the enzyme with the tRNA and were analysed in terms of the neutron scattering intensity at the origin $I(0)$ and of the radius of gyration as explained above. The results obtained can be summarized in the following points.

(a) The stoichiometry of the reaction is easily obtained from $I(0)$ for the data in H_2O, as explained above. Figure 2 shows the interaction of $tRNA_{Asp}$ from yeast with the corresponding synthetase. The data establish that two tRNA molecules bind to the dimeric enzyme, a point that was controversial with standard biochemical methods. The neutron method is very direct.

(b) The existence of electrostatic and specific interactions between the two molecules is, for instance, illustrated in figure 2. The interaction was studied in various media. The data in two of those media are shown in the figure. For molar ratio of tRNA:enzyme larger than two, the results are independent of ionic strength, whereas for smaller ratios much higher intensities are measured at low compared with high ionic strength. Similar behaviour has been observed with several systems (Zaccaï et al. 1979). The interpretation is unambiguous. At low ionic strength the tRNA induces the formation of aggregates of enzyme. Those aggregates are

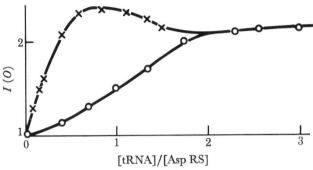

FIGURE 2. Titration of aspartic tRNA synthetase from yeast with the corresponding tRNA. On the vertical axis is shown the intensity at the origin normalized to unity for the enzyme alone. Buffer is 20 mM MES, pH 6.8 in H_2O, with 10 mM $MgCl_2$ (×) or 1.5 M $(NH_4)_2SO_4$ (○). (From R. Giege, D. Moras, J. C. Thierry and G. Zaccaï, unpublished work.)

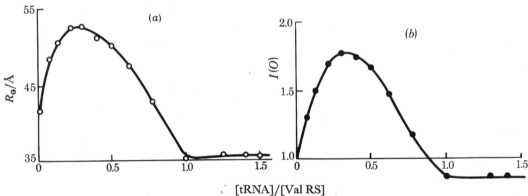

FIGURE 3. Titration of valyl tRNA synthetase from yeast with the corresponding tRNA: (a) the radius of gyration; (b) the intensity at the origin. Data are from buffer with 77% D_2O, with 50 mM potassium phosphate, pH 6.3.

dissociated either by adding tRNA or by increasing the salt content of the solution. Their electrostatic origin is obvious, from the sensitivity to ionic strength. Conversely, the stoichiometric complex obtained in the presence of excess tRNA and which is made through specific interactions is not sensitive to ionic strength. One may speculate that in the cell the electrostatic interaction favours the contact between the two molecules and makes easier the recognition between cognate molecules.

(c) The change of conformation of the enzyme is illustrated in figure 3a for $tRNA_{Val}$ from yeast (from Zaccaï et al. 1979). The only structural information so far has been deduced from the radius of gyration. The data shown are for solution in a solvent in which the tRNA is nearly invisible. At small tRNA:enzyme ratio, the increase of R_G corresponds to the above discussed aggregation. But when the stoichiometric complex is formed (in that case a 1:1 complex) the radius of gyration decreases from 40 to 35 Å. So the formation of the specific complex is accompanied by a huge change of configuration of the enzyme. Figure 3b shows that this change of conformation is accompanied by a decrease in the intensity at the origin, which is given by (2). As the enzyme concentration is unchanged, and the tRNA does not contribute to the scattering, there are only two possible explanations. The first would be a change of Σb through a change of deuteration of exchangeable protons associated with the change of conformation of the enyzme. This is excluded by the reversibility of the phenomenon. So the only remaining explanation is a change of specific volume of the system. So we therefore believe that we have established that, in appropriate conditions, the interaction between tRNA and the cognate tRNA synthetases is accompanied by a decrease by 1% of the specific volume. Our belief is that this is related to the modification of the hydration shells during interactions and the point is somewhat supported by the dependence of the phenomenon on ionic strength. This suggests that water may play a key role in protein–RNA interaction, and that the action of salt may be to some extent mediated through modification of the structure of the water in hydration shells.

Dr G. Zaccaï is thanked for communication of unpublished results.

REFERENCES (Jacrot)

Chauvin, C., Witz, J. & Jacrot, B. 1978 *J. molec. Biol.* **124**, 614–651.
Dessen, P., Blanquet, S., Zaccaï, G. & Jacrot, B. 1978 *J. molec. Biol.* **126**, 293–313.
Harrison, S. C., Olson, A. J., Schutt, C. E., Winkler, F. W. & Bricogne, G. 1978 *Nature, Lond.* **276**, 368–373.
Jacrot, B. 1976 *Rep. Prog. Phys.* **39**, 911–953.
Jacrot, B. & Zaccaï, G. 1980 *Biopolymers* (submitted).
Koch, M. H. J. & Stuhrmann, H. B. 1979 *Methods Enzymol.* **59**, 670–705.
Luzzati, V. 1979 In *Imaging processes and coherence in physics* (Lecture notes in physics), no. 212, pp. 207–215. Heidelberg: Springer-Verlag.
Serdyuk, I. N. 1979 *Methods Enzymol.* **59**, 750–775.
Serdyuk, I., Grenader, A. K. & Zaccaï, G. 1979 *J. molec. Biol.* **135**, 691–707.
Stuhrmann, H. G. 1974 *J. appl. Crystallogr.* **7**, 173–181.
Suck, D., Rayment, I., Johnson, J. E. & Rossmann, M. G. 1978 *Virology* **85**, 187–197.
Unge, T. & Strandberg, B. 1979 *Virology* **96**, 80–87.
Weidner, H., Yuan, R. & Crothers, D. M. 1977 *Nature, Lond.* **266**, 193–194.
Zaccaï, G., Morin, P., Jacrot, B., Moras, D., Thierry, J. C. & Giege, R. 1979 *J. molec. Biol.* **129**, 483–500.

Neutron diffraction from crystals of nucleosome core particles

By J. T. Finch†, A. Lewit-Bentley‡, G. A. Bentley‡,
M. Roth‡ and P. A. Timmins‡

† *M.R.C. Laboratory of Molecular Biology, Hills Road, Cambridge CB2 2QH, U.K.*
‡ *Institut Max von Laue – Paul Langevin, Avenue des Martyrs 156X,
38042 Grenoble Cedex, France*

The nucleosome is the basic repeating unit of chromatin (see review by Kornberg (1977)). It is a complex of histone protein molecules with a length of DNA, which digestion studies with the enzyme micrococcal nuclease have shown to be often about 200 base pairs in length but with quite wide variations between different cell species. With sufficient digestion, however, a 'core' particle is produced, which for all species so far investigated contains close to 145 base pairs of DNA associated with an octamer of pairs of the histones H3, H4, H2A and H2B. The molecular mass of the core particle is about 200000, roughly equally divided between DNA and protein.

From an earlier low resolution study by X-ray diffraction and electron microscopy on crystals of nucleosome core particles from rat liver chromatin, it was concluded that the core particle was flat, of dimensions about 110 Å × 110 Å × 57 Å†, somewhat wedge-shaped and strongly divided into two layers (Finch *et al.* 1977). A model was proposed in which the DNA was wound into about $1\frac{3}{4}$ turns of a flat superhelix of pitch about 28 Å around the histone octamer. The core particles in these crystals were found to have the histone proteins partly proteolysed, but their physico-chemical properties remained very similar to those of intact particles. Crystals have since been grown in the Cambridge laboratory from intact nucleosome cores. The unit cell is smaller than that of the proteolysed material, but is closely related to it; the space group is the same and the *a* and *b* unit cell dimensions very similar, but the *c* dimension is reduced by a factor close to 3. The approximate packing arrangement of the particle in this smaller cell was deduced from a low resolution, three-dimensional Patterson, calculated from the low angle X-ray data (Finch *et al.* 1978). This did not unambiguously resolve the relation of the particles to the screw axes of the space group, but it was found possible to choose signs for the strong low angle 0*kl* reflexions, which resulted in a Fourier map showing units with a very similar appearance to that in the corresponding map from the earlier crystals and preserving the local arrangement of units along the 2_1 screw axes parallel to *y* in that map.

The approximately equal division between protein and DNA makes the nucleosome core particle a good subject for study by neutron scattering by using the method of contrast variation. Several groups have made such a study on solutions of the particles (see, for example, Richards *et al.* 1977; Hjelm *et al.* 1977) and the Searle group in particular have shown that a model similar to that described above is consistent with their solution scattering data (Pardon *et al.* 1978). However, solution scattering suffers from the great disadvantage of not showing the orientation relation in the particle of the different features which give rise to the diffraction

† $1 \text{ Å} = 10^{-10} \text{ m} = 10^{-1} \text{ nm}$.

maxima. Moreover, the signal:noise ratio is very low for data at spacings smaller than about 35 Å. These disadvantages are largely overcome in crystal diffraction, and a low angle neutron diffraction study was therefore begun when sufficiently large crystals were produced.

The crystals of intact nucleosome cores have the space group $P2_1 2_1 2_1$, with unit cell dimensions $a = 111$ Å, $b = 198$ Å, $c = 111$ Å. The crystals used for this study had volumes 0.02–0.04 mm³. The data were measured at the I.L.L. on the small angle scattering camera D17, which has an area detector of 64 cm × 64 cm (128 × 128 counting elements) and which was adapted for single crystal work and software developed to collect integrated intensities. By using a wavelength of 9.2 Å, $h0l$ and $0kl$ sets of data were collected out to about $(25 \text{ Å})^{-1}$ for crystals soaked in mother liquors containing 0, 39, 65 and 90% D_2O. The 0% and 90% data sets required 5 days each and the intermediate sets 10 days each.

After correction for geometrical factors, the data from the different crystals were scaled as follows. At low angles of diffraction the reflections mainly relate to the particle shape, and their amplitudes depend on the difference between the scattering of the particle and that of the surrounding solution: there is a linear relation between the scattering amplitudes of centrosymmetric reflections and the concentration of D_2O. Approximate scaling factors were first deduced from a few reflections whose behaviours were clear, and more accurate factors calculated by linear regression by using all the data. The resulting scaled amplitudes showed a good linear dependence on D_2O concentration. This method of scaling is independent of the absolute signs of the reflections at any one contrast (D_2O concentration), but if the signs for one contrast are known, the signs of the reflections at all other contrasts can be deduced from the linear relation between structure factor and contrast.

The $0kl$ data have been used most extensively for sign determination so far, since it is known from the X-ray work that the particles in the crystal are arranged in approximately hexagonally packed columns, and seen in the direction of the x-axis of the crystal there is little overlap between the projections of neighbouring columns (Finch *et al.* 1978); this projection thus yields the clearest picture of individual core particles. As implied above, the problem in the analysis is to get a starting set of signs for one contrast: for this we turned to the X-ray work. For the crystal in 0% D_2O, the relative scattering of neutrons by protein and DNA should not be all that different from that of X-rays. Thus the signs used to calculate the X-ray Fourier map (Finch *et al.* 1978) were used to calculate the corresponding neutron map in 0% D_2O, and were changed as indicated by the contrast variation series to calculate the maps for other contrasts. Of the resulting maps, that corresponding to the 39% D_2O data was particularly disappointing. At this concentration, the scattering of D_2O is close to that of protein and so the diffraction pattern is mainly due to DNA for which there was the well defined model described above. However, the positive density in that map was not continuous and did not always overlap positive density in the 0% map, and there were also unrealistically strong negative regions. A trial-and-error method was therefore adopted and by this it was found possible to choose signs for the strong reflections of the 39% data that produced a Fourier map strikingly similar to the projection of about $1\tfrac{3}{4}$ turns of a superhelix (figure 1).

At a concentration of 65%, the scattering density of D_2O closely matches that of DNA and thus the crystal scattering is dominated by that of the protein. By using the signs corresponding to this data set predicted from those used in figure 1 together with other strong reflections of appropriate sign, the Fourier map shown in figure 2c was calculated. It shows the expected columns of protein, and the heart shape is very similar to a projection of the three-dimensional

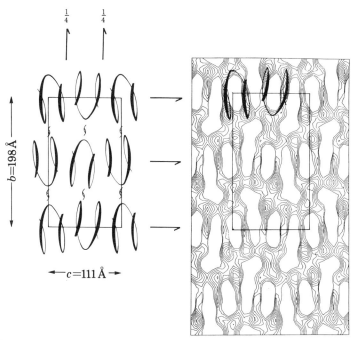

FIGURE 1. A Fourier map (right) calculated from the $0kl$ neutron scattering data from a nucleosome core crystal soaked in 39 % D_2O. At this concentration, the scattering of D_2O closely matches that of protein and so virtually the DNA alone is seen in contrast. With the signs chosen for the reflexions here, the positive density in the map correlates strikingly with the projection of about $1\frac{3}{4}$ turns of a superhelix arranged as indicated in the diagram of the x-projection of the unit cell on the left.

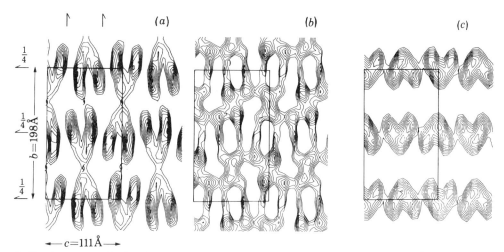

FIGURE 2. Fourier maps calculated from $0kl$ neutron scattering data to a resolution of 25 Å, from crystals soaked in (a) 0 % D_2O, in which both DNA and protein are contrasted relative to the mother liquor, (b) 39 % D_2O, in which DNA shows dominant contrast, and (c) 65 % D_2O in which protein is dominantly contrasted. The structure factors used in (b) (figure 1) were changed in amplitude and sign as indicated by the contrast variation series for calculation of the maps in (a) and (c). The three maps show the self-consistency of the sign combination used, and indicate a feasible relation between the histone octamer (c) and the DNA superhelix (b). The map at 90 % D_2O is not shown but is roughly midway between maps (a) and (c), as one would expect from the relative contrasts of protein and DNA at this D_2O concentration.

[157]

image reconstruction from electron micrographs of the isolated histone octamer carried out in Cambridge (Klug *et al.* 1980). Furthermore, this reconstruction has a twofold axis which in the Fourier map lies a few degress away from the direction of the y-axis, more or less coincident with the twofold axis of the corresponding DNA superhelix (figure 1). Thus, to this resolution at least, the core particle has an overall twofold axis.

The corresponding map at 0 % D_2O (figure 2a) shows units with a similar wedge-shaped, bifurcated appearance to that obtained in the earlier X-ray work from the large unit cell (cf. fig. 6 of Finch *et al.* 1977). The positions of the units with respect to each other are very similar to those in the earlier interpretation of the X-ray data from the small cell (Finch *et al.* 1978) but they are located differently with respect to the screw axes of the unit cell. The distinction between these two arrangements was not obvious in the three-dimensional Patterson map of the low resolution data. However, more recent Patterson maps, which include higher angle X-ray data, clearly favour the arrangement found here and also confirm the existence of the particle dyad lying a few degress from the y-axis (Finch *et al.* 1980).

Although one must always bear in mind the dangers of the trial-and-error method, the results described above do give confidence in the signs adopted for the neutron scattering data for this projection. They result in a set of Fourier maps which are self consistent for all contrasts. Moreover, the maps at 39 % and 65 % D_2O show a striking correlation both with the DNA superhelix of the proposed model and also with the histone octamer structure obtained from the three-dimensional image reconstruction which is not only consistent with the twofold symmetry of each. The result also defines the relative orientations of the two components about the common dyad in a way which in three dimensions is highly plausible (Klug *et al.* 1980). Thus, in addition to confirming the model, the neutron data provide a basis for further interpretation of the interaction between the protein and the DNA.

The participation of J. T. F. was financed by the Science Research Council.

References (Finch *et al.*)

Finch, J. T., Brown, R. S., Rhodes, D., Richmond, T., Rushton, B., Lutter, L. C. & Klug, A. 1980 *J. molec. Biol.* (Submitted.)

Finch, J. T., Lutter, L. C., Rhodes, D., Brown, R. S., Rushton, B., Levitt, M. & Klug, A. 1977 *Nature, Lond.* **269**, 29.

Finch, J. T., Lutter, L. C., Rhodes, D., Brown, R. S., Rushton, B. & Klug, A. 1978 In *12th FEBS Meeting, Dresden*, vol. 51, p. 193.

Hjelm, R. P., Kneale, G. G., Suau, P., Baldwin, J. P. & Bradbury, E. M. 1977 *Cell* **10**, 139.

Klug, A., Rhodes, D., Smith, J., Finch, J. T. & Thomas, J. O. 1980 *Nature, Lond.* (Submitted.)

Kornberg, R. D. 1977 *A. Rev. Biochem.* **46**, 931.

Pardon, J. F., Cotter, R. I., Lilley, D. H. J., Worcester, D. L., Campbell, A. H., Wooley, J. C. & Richards, B. H. 1978 *Cold Spring Harb. Symp. quant. Biol.* **42**, 11.

Richards, B., Pardon, J. F., Lilley, D., Cotter, R. & Worcester, D. 1977 *Cell Biol. int. Rep.* **1**, 107.

Molecular dynamics of hydrated proteins

By H. D. Middendorf[†] and Sir John Randall, F.R.S.[‡]

[†] *Department of Biophysics, University of London King's College, London WC2B 5RL, U.K.*

[‡] *Department of Zoology, University of Edinburgh, West Mains Road, Edinburgh EH9 3JT, U.K.*

This contribution presents the first results of an application of high-resolution quasi-elastic neutron scattering to the dynamics of protein hydration. Using two samples (film stacks and 'fluffy' powders) of biosynthetically fully deuterated C-phycocyanin extracted from blue-green algae, we have studied the Doppler-like broadening of the quasi-elastic line as a function of scattering angle 2θ at several sub-monolayer H_2O hydration levels. The backscattering spectrometer IN 10 at the Institut Laue–Langevin (I.L.L.), Grenoble, was employed to measure wet-minus-dry difference broadenings ΔE of up to 5×10^{-3} cm^{-1}, or 0.4 µeV, for momentum transfers $k = (4\pi/\lambda_0) \sin \theta$ between 0.15 and 1.7 Å^{-1} (incident wavelength $\lambda_0 = 6$ Å). The results show that $\Delta E(k)$ possesses an oscillatory structure with a first maximum between $k_{max} = 0.4$ and 0.8 Å^{-1}. The position of this maximum shifts to higher k with increasing hydration, while its intensity increases and the following minimum at $k_{min} = 0.7$–1.3 Å^{-1} becomes progressively more shallow. Structure factor measurements indicate that line narrowing due to structure in $S(k)$ is not the dominant mechanism in determining this oscillatory behaviour of $\Delta E(k)$. A Chudley–Elliott jump diffusion model was adopted as a working hypothesis to extract a characteristic length (water migration distance a) and a characteristic time (residence time τ_0 at a hydration site) from the $\Delta E(k)$ data. Values of $a = 6$–9 Å and $\tau_0 = 5$–30 ns were obtained for the powder sample and are shown to agree well with the average jump distances derived from topographical considerations in conjunction with hydration number estimates.

1. Introduction

Neutron inelastic scattering was first used to study the hydration dynamics of biopolymers about 10 years ago (Whittemore 1968; Dahlborg & Rupprecht 1971). It had been realized for some time that radiation scattering techniques employing cold neutrons could contribute significantly to the understanding of hydration processes in complex organic materials, both because of the unique property of neutrons to resolve structural *and* dynamical detail at the molecular level, and because of the large scattering contrast between hydrogen and deuterium. Only during the past few years, with the advent of new high-resolution spectrometers on sources of greatly increased flux, has it become possible to realize some of the early expectations.

In this contribution we present the first results of an application of high-resolution quasi-elastic neutron scattering to the dynamics of protein hydration. Our procedure in this work has been to start at the low-hydration end of the sorption isotherm, i.e. to use moist powders and slightly hydrated film samples rather than proteins in solution. This was done in order first to establish the practice and interpretation of hydration difference experiments from biosynthetically deuterated macromolecules for which there is a large coherent component in the scattering, and also to make better use of the very limited instrument time available to us. A preliminary account of this work has already been given (Randall *et al.* 1978). An earlier, exploratory experiment with a low-resolution spectrometer was carried out at A.E.R.E. Harwell (J. T. Randall & S. Gilmour, unpublished observations, 1975).

2. Protein hydration and dynamics

Water is an important and ubiquitous constituent of all living organisms. Many structural and functional aspects of its interaction with biological molecules have been studied by a variety of techniques ranging from calorimetry to nuclear magnetic resonance (Kuntz & Kauzmann 1974; Berendsen 1975; Packer 1976; Hopfinger 1977; Edsall & McKenzie 1978; Mathur-De Vré 1979). The thermodynamic information on protein hydration processes is provided by measurements of the entropy and enthalpy of sorption. The isopiestic method of Bull (1944; Bull & Breese 1968) is frequently used to measure sorption isotherms, and for proteins and polypeptides these are all of sigmoidal type II (Brunauer *et al.* 1938). At very low relative humidities (r.h.) the water uptake of globular proteins shows a steep rise until at *ca.* 0.40–0.80 g H_2O/g protein the primary hydration sites are occupied. Following this there is a 'knee' in the sorption isotherm and further increases in r.h. level lead to a steady but slower increase in water uptake.

While a limited amount of *structural* information on the water of hydration in and around protein molecules can be inferred from isotope exchange studies, infrared spectroscopy and nuclear magnetic resonance, much more detailed information comes from X-ray and neutron diffraction experiments on protein crystals. These have shown, in particular, that a small number of water molecules (3–25) are tightly held in the interior of the protein and that these must be regarded as an essential part of its tertiary and quaternary structure. A larger number (up to several hundred) are less tightly bound but still localizable at or near the molecular surface.

The currently available *dynamical* information on protein hydration at the molecular level comes from a number of spectroscopic techniques sensitive in different regions of the spectrum and to different properties of the water molecule and its immediate environment. These provide measurements mainly of the characteristic times τ_r for rotational diffusion or reorientational 'tumbling'. The results show, broadly, that there are three classes of water molecules close to a protein in solution: first, a small number bound almost irrotationally ($\tau_r = 10^{-5}$ to 10^{-7} s) at specific hydration sites offering very favourable multiple hydrogen-bonding conditions; secondly, a larger number (comparable to monolayer coverage) interacting less strongly with the protein but still sufficiently restricted motionally so that $\tau_r \approx 10^{-9}$ s; thirdly, all water molecules beyond one or two monolayers that possess dynamical properties essentially indistinguishable from bulk water ($\tau_r = 10^{-10}$ to 10^{-11} s). Because of the heterogeneity of the protein surface both with respect to topography and binding characteristics (Nedev & Churgin 1975), the distinctions made here are not sharp and depend very much on interpretational details of the techniques used.

Another approach to the study of hydration processes and protein dynamics in general is by numerical simulation. With the increasing availability of large and fast computers, investigations on the static accessibility of a protein to solvent molecules by space-filling techniques (Finney 1977; Lee & Richards 1971) are being extended to full-scale simulations of the dynamics of the water of hydration (Hermans & Rahman 1976; Clementi *et al.* 1977, 1979;

Hagler & Moult 1978). The internal motions in crystallographically known proteins are being studied on an atomic scale by *ab initio* molecular dynamics simulations in the time domain 10^{-12} to 10^{-10} s (Levitt 1976; McCammon *et al.* 1977). Further theoretical work of interest in this context is directed towards an understanding of the conformational flexibility and transient accessibility of molecular groups in proteins (Karplus & Weaver 1976; Cooper 1976).

It is apparent from many of these studies that there is a relative lack of experimental information on the energy states of proteins in the nanosecond to picosecond region (Williams 1978). Neutron spectroscopic methods are in principle capable of bridging this region by overlapping with nuclear magnetic resonance techniques on the low-frequency side, and with optical techniques on the high-frequency side.

3. Basic aspects of neutron scattering

Quasi-elastic neutron scattering explores the low-frequency region of the spectrum characterized by energy transfers of the order of, or smaller than, 1 cm^{-1}†. In this region, monochromatic neutron pulses scattered by a sample at ordinary temperatures undergo Doppler-like broadening by non-quantized diffusive motions of various kinds and by a quasi-continuum of slow rotational modes that merges into or overlaps with the inelastic spectrum proper. Because of the large incoherent cross section of protons compared with all other nuclei (Bacon 1962), the scattering from *natural* biological samples is almost wholly incoherent and the hydrogen atoms therefore act as probes of the molecular dynamics. This changes drastically on deuteration, and in the limiting case of a completely deuterated sample the scattering will be predominantly coherent. The technique of H–D contrast variation, already well established in neutron diffraction (Worcester 1976; Jacrot 1976; Kneale *et al.* 1977; Stuhrmann & Miller 1978), is of equal importance in quasi-elastic and inelastic scattering as it provides a means to weight the cross section of a heterogeneous system in favour of a particular component (Egelstaff 1975; White 1975).

In a quasi-elastic or inelastic scattering experiment the double differential cross section, $d^2\delta/d\Omega dE$, is measured with respect to solid angle Ω and energy E of scattered neutrons. Momentum transfer $\hbar k$ and energy transfer $\hbar\omega$ of a neutron deflected upon scattering by an angle 2θ are obtained from the conservation equations

$$\hbar \boldsymbol{k} = \hbar(\boldsymbol{s}-\boldsymbol{s}_0); \quad \cos 2\theta = \boldsymbol{s}\cdot\boldsymbol{s}_0/ss_0; \tag{3.1}$$

$$\hbar\omega = E - E_0 = \tfrac{1}{2}m(v^2 - v_0^2); \tag{3.2}$$

where $\boldsymbol{s}, |\boldsymbol{s}| = s = 2\pi/\lambda$, v and E are the wavevector, wavenumber, wavelength, velocity and energy of the scattered neutron, respectively, and subscript zero denotes incident beam quantities (m is the neutron mass, $\hbar = h/2\pi$). For $\omega = 0$ (no energy exchange), these equations reduce to the familiar condition $k = (4\pi/\lambda_0)\sin\theta$ for elastic scattering. In quasi-elastic scattering, the energy change of a neutron upon scattering is small compared with the incident energy E_0. For $\lambda_0 \gtrsim 5$ Å, the wavelengths normally used in studies of this kind, E_0 is smaller than the thermal energy $k_B T$ by a factor of at least 8. The parameter range of interest here is thus characterized by the inequalities

$$|\hbar\omega| \ll E_0 \ll k_B T. \tag{3.3}$$

† 1 cm^{-1} ≡ 124 μeV.

The connection between $d^2\delta/d\Omega dE$ and the molecular dynamics of the sample is made by correlation functions describing the evolution of the probability distribution of its nuclei in space and time (Glauber 1962; Egelstaff 1967). The differential cross section can always be written as the sum of an incoherent and a coherent part; for an assembly of N nuclei of the same species this is

$$\frac{d^2\delta}{d\Omega dE} = \frac{N}{\hbar}\frac{s}{s_0}[b_{inc}^2 S_{inc}(\boldsymbol{k}, \omega) + b_{coh}^2 S_{coh}(\boldsymbol{k}, \omega)], \qquad (3.4)$$

where
$$S_{inc}(\boldsymbol{k}, \omega) = \frac{1}{2\pi}\iint G_s(\boldsymbol{r}, t)\, e^{-i(\boldsymbol{k}\cdot\boldsymbol{r}-\omega t)}\, d\boldsymbol{r}\, dt$$

and
$$S_{coh}(\boldsymbol{k}, \omega) = \frac{1}{2\pi}\iint G(\boldsymbol{r}, t)\, e^{-i(\boldsymbol{k}\cdot\boldsymbol{r}-\omega t)}\, d\boldsymbol{r}\, dt.$$

Here G_s and G are van Hove's self-correlation and pair correlation function, respectively, and b_{inc}, b_{coh} are the incoherent and coherent scattering lengths. The factorization of the two parts in (3.4) into a cross section $\propto b^2$ and a scattering law $S(\boldsymbol{k}, \omega)$ is possible only for a simple system. For a heterogeneous molecular sample, the incoherent part of (3.4) must be replaced by appropriate sums over atomic species or groups and their particular scattering laws, but the formulation of this is still relatively straightforward. The coherent part, however, because it describes the interference effects, assumes a much more complex form. The theory of inelastic neutron scattering from molecular systems has been developed by Steele & Pecora (1965), Larsson (1971) and Egelstaff et al. (1975).

4. Samples

Samples of the biosynthetically deuterated protein used in this study, C-phycocyanin, were initially provided by D. S. Berns, Albany Medical College, New York, and subsequently by H. L. Crespi, Argonne National Laboratory, Illinois. Dr Berns's material had been extracted from the blue-green alga *Plectonema calothricoides* and that of Dr Crespi from *Synechococcus lividus* grown in 99.7% pure D_2O (Taecker et al. 1971; Crespi 1977). Blue-green algae contain large amounts (up to 25%) of phycocyanin and are capable of synthesizing it in extreme environmental conditions. Phycocyanin is a multimeric chromoprotein (molecular mass 190 000) located in the stroma region between the thylakoid membranes, and its function is that of a light-harvesting protein involved in photosynthesis as part of photosystem II (O'Carra & O'hEocha 1976; Rüdiger 1980). Although its crystallographic structure is not known, its biophysical and biochemical properties have been studied since the days of Svedberg. Current interest in protio- and deuterio-phycocyanin relates primarily to its photochemical properties and to the opportunities it offers as a model system for studying the inter- and intramolecular forces involved in protein folding and aggregation processes (Berns 1971).

The phycocyanin samples were prepared in two forms. One consisted of glassy protein films cast on 9 μm pure Al foils (surface density *ca.* 5 mg cm^{-2}); this was used mainly in our initial experiments on IN 10 (see below). The other was in the form of fluffy aggregates of wisp-like particles obtained by drying from aqueous solutions, placed loosely on a stack of $4 \times 5 \times 40$ mm trays made from 9 μm Al foil. This method of preparation was used in all low-hydration experiments on IN 10 and IN 5. It made it possible, because of the greatly increased surface area, to re-equilibrate the sample to a new, sufficiently close hydration level during a reasonable time interval, without disturbing its geometry. A cylindrically symmetrical version of this sample holder was used for the powder diffraction experiment on D1B; here the stack of circular trays was made from 25 µm V foil. The total amount of protein in the beam was between 300 and 500 mg.

The samples were hydrated inside a temperature and humidity controlled sample changer equipped with a movable axial rod carrying an Al frame with the sample holder, a thermocouple and a solid-state humidity sensor. Dry helium of 99.9% purity was passed through wash bottles containing distilled H_2O or saturated salt solutions in H_2O; these bottles were kept at a controlled temperature different from that of the sample changer and connected with the sample changer by a heatable steel pipe. The humidity level as measured by an M.C.M. Model 700L hygrometer was recorded continuously, together with the total count rate of scattered neutrons. The humidity control system described here was gradually implemented during four experiments; our earlier measurements were less quantitative but these are currently being recalibrated. The 'dry' reference state of a sample was defined by purging with dry helium for 10–20 h until the lag between neutron count rate and hygrometer reading disappeared and both traces had reached an essentially flat asymptote. Temperatures were controlled to within ± 0.5 °C.

5. INSTRUMENTATION

Three neutron scattering instruments at the Institut Laue–Langevin, Grenoble, were used in this study: The backscattering spectrometer IN 10 (Birr *et al.* 1971), the multi-chopper time-of-flight spectrometer IN 5 (Lechner *et al.* 1973), and the powder diffractometer D1B (Convert 1975).

The results discussed in this paper are mainly from experiments performed on IN 10. The backscattering technique achieves a quasi-elastic resolution of 0.003 to 0.01 cm^{-1} (f.w.h.m.) as follows. A spatially collimated beam of neutrons from the cold source in the reactor (wavelength distribution 2–12 Å†) is reflected from a silicon monochromator crystal, which gives a highly monochromatic line of neutrons with $\lambda_0 = 6.2708$ Å. By imparting a periodic translational motion of a few hertz to the monochromator, the neutrons selected by it are Doppler-shifted with respect to the fixed sample and continuously scan a spectral window of 0.045 Å in wavelength or 0.24 cm^{-1} in energy centred on λ_0. On scattering from the nuclei in the sample, these neutrons suffer small energy changes characteristic of thermal and collective motions. The scattered neutrons are energy-analysed by 180° reflexion from a fixed array of silicon crystals positioned 1.5 m from the sample at the scattering angle 2θ observed. The crystals forming this array reflect neutrons back on to a detector located close to the sample. Five to seven detectors are arranged on a semicircle of 5 cm radius concentric with the sample changer axis, opposite the array of analyser crystals. A large cylindrical segment of the sample changer in this region consists of 12 µm pure Al foil and is transparent to cold neutrons. The total Al thickness traversed by the primary beam or the scattered neutrons nowhere exceeds 60 µm. The detectors were distributed over scattering angles from $2\theta = 8.6$ to 114°, corresponding to $k = 0.15$ to 1.68 Å^{-1}.

On IN 5, a conventional time-of-flight spectrometer of high resolution, the incident wavelength was 10 Å and eight detector banks were distributed over the scattering angle range $2\theta = 10$ to 130° ($k = 0.11$ to 1.14 Å^{-1}). The sample changer used in this experiment was

† 1 Å = 10^{-10} m = 10^{-1} nm.

identical in design except for a larger diameter of 12 cm. On the powder diffractometer D1B, the incident wavelength was 2.52 Å and the multi-detector consisted of 400 cells covering the range $2\theta = 4$ to $84°$ ($k = 0.17$ to 3.33 Å^{-1}). Here the sample container was a 2.5 cm diameter V cylinder. The standard I.L.L. user's programs available for each instrument were used to monitor-normalize the raw data and to apply corrections for detector efficiency.

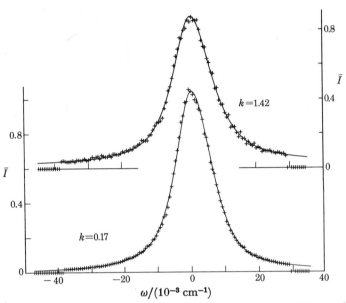

FIGURE 1. Examples of quasi-elastic spectra measured (+) for d-phycocyanin film stacks hydrated at relative humidity r.h. $\lesssim 90\%$ H$_2$O (temperature 24 °C). I = neutron counts corrected for detector efficiency (arbitrary units); ω = energy transfer. Momentum transfer k (Å^{-1}) corresponds to scattering angles $2\theta = 9.8$ and $90°$; resolution widths W_r (dry sample) were 0.0129 and 0.0143 cm^{-1}, respectively. Solid lines are analytical fits obtained by using asymmetric Peyre–Principi profiles. The term d-phycocyanin will henceforward be used to indicate the fully deuterated protein.

6. EXPERIMENTAL RESULTS

(a) IN 10

In our first experiment we used an exchange-protonated film stack sample of d-phycocyanin and recorded sets of quasi-elastic spectra for a sequence of H$_2$O hydration runs up to approximately 90% relative humidity (r.h.), starting with an almost dry sample (spectral window 0.072 cm^{-1}, six detectors). Two examples for the line shapes thus obtained are shown in figure 1. Conventional line broadening analysis (single-Lorentzian fitting) with respect to the reference or resolution widths measured for the dry sample and a Perspex scattering standard gave small broadenings ΔE of up to 3.2×10^{-3} cm^{-1} (f.w.h.m.), or 25% of the resolution width. The unexpected finding was that these broadenings showed a maximum at the position of the 4th detector ($k = 0.85$ Å^{-1}) together with the hint of an oscillatory structure appearing at higher angles. Subsequent experiments designed to check this result and guard against possible geometrical or instrumental effects have confirmed unambiguously that $\Delta E(k)$ indeed possesses an oscillatory structure in the hydration range investigated so far. This was done by changing the orientation of the sample with respect to the incident beam (45, 68 and 90°), by employing a different sample holder (perforated Al foil pouch instead of tray stack), changing the spectral window (background test) and interchanging polished and unpolished analyser crystals (different resolution).

In order to study in greater detail the low-hydration régime and to create more uniform hydration conditions by increasing the effective surface area, the 'fluffy' powder samples described in §4 were used in all further experiments. Also, to obtain reliable values for quasi-elastic broadenings that could be ascribed to changes in hydration alone, the spectra recorded for a given hydration level of the sample were analysed not with respect to the instrumental resolution width W_v of the quasi-elastic line (as measured for a purely incoherent scattering standard), but differentially with respect to the width W_r obtained for the identical, dry or very slightly hydrated sample. Under conditions of undisturbed sample and scattering geometry during a series of hydration runs, and sufficiently good counting statistics (more than a few thousand counts integrated over the channels within the half-width points), it was possible to determine relative broadenings *for a given detector* with an estimated error of $\pm (1.5-2) \times 10^{-4}$ cm^{-1}. Because of the large differences in resolution between individual detectors, and line shape distortions due to the impossibility of achieving uniform backscattering conditions for all six or seven detectors, the error in ΔE with respect to scattering angle (at a fixed hydration level) was larger by a factor of perhaps 2.

Considerable care needed to be exercised also in the data analysis. Standard single-Lorentzian fitting routines were found to be inadequate because of the smallness of the broadenings and appreciable non-Lorentzian intensity increases in the wings of the lines. Least-mean-square fits by superpositions or convolutions of Gaussian and Lorentzian functions were used instead. In one method, two parabolic sections meeting at the peak were first fitted to the data points in the upper 25–40 % of a line (depending on the number of points and their scatter) to define a centroid. Asymmetric Peyre–Principi profiles (Middendorf 1974) with a flat background were then fitted separately to the two halves of the line. The functional form used for these was

$$I_p(\omega) \approx a_0 + pL(\omega/w_0) + (1-p)\, G(\omega/w_0), \tag{6.1}$$

where
$$L(x) = 1/(1+x^2); \quad G(x) = \exp(-x^2 \ln 2) \tag{6.2}$$

and w_0 is the half width at half maximum when $a_0 = 0$ (no background). Similar fits with the use of the convolution product $L(\omega/w_l) \otimes G(\omega/w_g)$ instead of (6.1) gave analytical approximations that were essentially indistinguishable from the Peyre–Principi (P.P.) profiles, and the latter were therefore employed in most of the data analysis. Two such profiles are shown in figure 1. These analytical representations were then used to calculate contour level widths in steps of 5 % between 5 % and 80 % of the peak intensity with respect to a_0 as the baseline. In addition to the 50 % widths thus obtained (denoted ΔE), an attempt was made to define a measure of the broadening (denoted ΔE^*) based on an integral rather than a local property of the quasi-elastic line and therefore more sensitive to non-Lorentzian line shape changes. The peak-normalized lines were integrated over ω within the spectral window of the instrument and corrected for fractional wing intensities outside the window by using the fitted P.P. profiles, to give corrected total intensities $\bar{A}_r$ and $\bar{A}_s$ for the reference line (dry sample) and the broadened line (wet sample), respectively. Then $\Delta E^* = \bar{A}_s - \bar{A}_r$ and this is related to the analytical approximation (6.1) by

$$\Delta E^* = f(p)\, w_{s_0} - f(q)\, w_{r_0}; \quad f(x) = \pi x + (1-x)(\pi/\ln 2)^{\frac{1}{2}}, \tag{6.3}$$

where w_{s_0}, w_{r_0} and p, q are the widths (h.w.h.m.) and P.P. parameters of sample and reference lines. For purely Lorentzian profiles, $\Delta E^* = \frac{1}{2}\pi \Delta E_{\text{Lorentz}}$. Comparison with Lorentzian-fitted lines gave values between 1.7 and 2.0 for the proportionality factor appearing here, instead of $\frac{1}{2}\pi$. Background corrections were determined from those measurements for which the wider spectral window of the instrument was employed; these ranged from 0 to 4.5 % of the peak intensity. The principal conclusions about $\Delta E(k)$ and its dependence on hydration proved to be unaffected by background corrections of this kind.

For the 'fluffy' sample configuration at low hydration levels, the variation of the quasi-elastic broadening with scattering angle was found to be more pronounced and the maximum occurred at a lower angle. In the first experiment a proton-exchanged sample was hydrated with H_2O continuously from the dry state at an extremely low rate for 40.53 h (constant inlet flow of very slightly humidified helium). The uptake of water by the sample as reflected in the total detector count rate is shown in figure 2. The first set of spectra was taken after 5.45 h; the initial steep rise in the count rate is due to the rapid hydration of primary surface sites. The broadenings ΔE^* from this experiment, as evaluated from spectra accumulated over two intervals of several hours, are shown in figure 3 and compared with those obtained for the film stack sample described above.

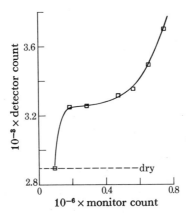

FIGURE 2. Total count rate of scattered neutrons plotted against direct beam monitor count (a count of 18 600 represents 1 h) for 'fluffy' powder sample of d-phycocyanin under conditions of continuous hydration (instrument IN 10; $T = 25\ °C$).

Whereas this experiment was deliberately performed under slowly changing hydration conditions, an attempt was made in subsequent experiments with the use of 'fluffy' powder samples, following improvements in hygrometry instrumentation, to achieve a reasonable degree of quasi-equilibrium during extended periods of stepwise hydration. The results are shown in figure 4. It is seen that there is a systematic shift in the position of the maximum in $\Delta E(k)$ with increasing hydration, while the minimum at $k = 0.7$ to 1.1 becomes progressively more shallow.

(b) IN 5

Fully instrumented experiments performed in parallel with those on IN 10 and covering the 0.1–500 cm^{-1} energy transfer region have only recently begun. The quantity of immediate interest in connection with the interpretation of IN 10 data is the elastic incoherent structure factor (Springer 1972; Volino & Dianoux 1978) which, in favourable circumstances, enables the translational and rotational contributions to the quasi-elastic intensity to be separated. Three examples of the spectra obtained for an H_2O-hydrated powder sample at $2\theta = 54$, 96 and 133° are shown in figure 5. Also shown is the spectrum of the dry sample at $2\theta = 54°$;

all dry sample spectra, apart from small effects in the wings, were similar to this and gave no detectable broadening, i.e. closely reproduced the instrumental resolution function. The main qualitative result for the hydrated sample is that the central elastic peak is essentially unbroadened (at a resolution of 0.17 cm^{-1}) throughout the k range covered so far, and that it is superimposed on a relatively broad Lorentzian-like feature the intensity of which increases with k, its width being of the order of 1 cm^{-1}. The separation of these two features does not present any difficulty, and we expect to be able to determine the elastic incoherent structure factor from current and future IN 5 experiments as required for a full interpretation of the quasi-elastic scattering.

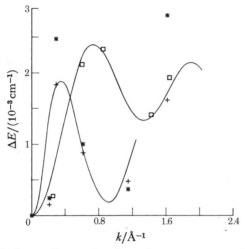

FIGURE 3. Quasi-elastic broadenings ΔE as a function of momentum transfer k for film stack sample (□) (hydration conditions as in figure 1), 'fluffy' sample at low (+) and intermediate (∗) hydration levels (r.h. = 10–30% and 30–50%, respectively).

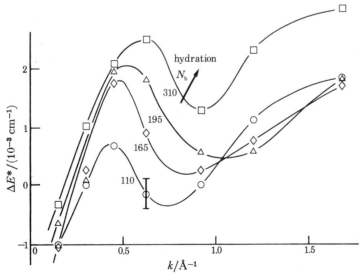

FIGURE 4. Hydration difference broadenings $\Delta E(k)$ for 'fluffy' powder sample at four hydration levels corresponding to r.h. conditions defined by saturated solutions of LiCl (r.h. ca. 15%), NaCl (ca. 30%), NaHSO$_4$ (ca. 52%) and (NH$_4$)$_2$SO$_4$ (ca. 81%) in H$_2$O at 24 °C. Solid lines drawn in to guide the eye.

(c) D1B

The purpose of this diffraction experiment was twofold: to determine the structure factor $S(k)$ for the dry powder sample of phycocyanin and for several hydration levels under conditions essentially identical with those on IN 10 and IN 5, and to look for possible conformational changes associated with the transition from the dry to a very slightly hydrated state, and further to the first two or three hydration levels studied on IN 10 and IN 5. Only the first of these objectives will be considered in connection with the interpretation of $\Delta E(k)$ discussed in §7.

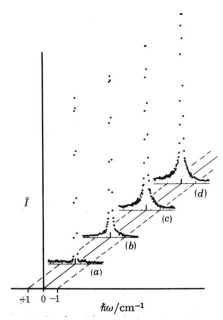

FIGURE 5. Examples of quasi-elastic spectra obtained on IN 5 for (a) dry d-phycocyanin sample at $2\theta = 54°$; (b), (c), (d) hydrated sample (r.h. ca. 30 %, NaCl in H_2O) for $2\theta = 54, 96$ and $133°$, respectively. Normalized to give approximately equal peak heights.

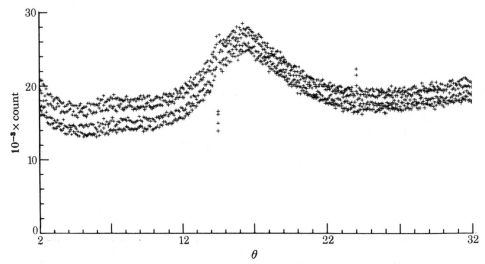

FIGURE 6. Structure factor data obtained by using powder diffractometer D1B for dry d-phycocyanin and for three hydration levels essentially identical with those used on IN 10. The traces shown are the result of subtracting the pattern obtained for a blank container run from those recorded for sample runs.

The results for a proton-exchanged sample hydrated in steps according to the procedure described before are shown in figure 6.

7. Interpretation of $\Delta E(k)$

The quantities of primary interest in quasi-elastic scattering studies are the total intensity $S(k)$ and the half-width $\Delta E(k)$ of the quasi-elastic peak. The dependence of these quantities on H–D contrast and temperature can give important clues to the nature of the process observed. Further information may be extracted from a detailed line shape analysis of the quasi-elastic scattering as a function of k. In the present paper we shall discuss only $\Delta E(k)$ and, to a limited extent, $S(k)$.

The significance of k is that the motional properties of the sample are probed over a scale length of about $2\pi/k$ (Å). Thus, in incoherent scattering, the macroscopic limit of classical diffusion is observed as $k \to 0$, and the behaviour of $\Delta E_{\text{inc}}(k)$ for $k \gtrsim 1$ reflects the medium and short-range motions over molecular and atomic distances. The detailed form of $\Delta E_{\text{inc}}(k)$ has been investigated both theoretically and experimentally for a variety of dynamic processes (Vineyard 1958; Egelstaff 1967; Springer 1972). The simplest case is that of continuous Fick's law diffusion for which $\Delta E_{\text{inc}} = 2\hbar D k^2$, where D is the self-diffusion coefficient. In hydrogenous molecular liquids, rotational effects cause departures from Fick's law behaviour such that $\Delta E_{\text{inc}}(k^2)$ for $k \to 1$ falls below the straight line defined by the $k \to 0$ asymptote but always increases monotonically with k. Oscillatory behaviour of $\Delta E_{\text{inc}}(k)$, on the other hand, is characteristic of a class of jump diffusion processes discussed widely in the context of neutron scattering from protons diffusing on interstitial lattice sites in metal hydrides (Sköld & Nelin 1967; Gissler & Rother 1970; Rowe et al. 1971). For an ensemble of incoherently scattering particles migrating on a two-dimensional or three-dimensional lattice of sites, model calculations indicate that line-width changes similar to those shown in figures 3 and 4 occur whenever the residence time τ_0 of a particle at a site is large compared with the time during which it moves to a neighbouring site. Because of the difficulty of formulating the space and time dependent pair correlation function, $G(r, t)$, for a coherently scattering molecular system of complex structure (compare §3), measurements of the line broadening $\Delta E_{\text{coh}}(k)$ have been analysed in detail only for a few cases. Those on polymers (Allen & Higgins 1973; Maconnachie & Richards 1978) are of some relevance here. Fundamental relations between the coherent and incoherent parts of the scattering law and their moments with respect to ω (de Gennes 1959; Sköld 1967) indicate that one should expect a line narrowing, i.e. a dip in $\Delta E(k)$, in those regions of k where the structure factor $S(k)$ possesses a maximum, and vice versa. In the classical limit, all moments up to the second are satisfied by an expression known as the effective mass approximation (e.m.a.) and given by

$$S_{\text{coh}}(k, \omega) = S(k)\, S_{\text{inc}}[k/\sqrt{\{S(k)\}}, \omega]. \tag{7.1}$$

This model leads to
$$\Delta E_{\text{coh}}(k) = \Delta E_{\text{inc}}[k/\sqrt{\{S(k)\}}]. \tag{7.2}$$

Turning now to the interpretation of the data presented in §6, we note, first of all, that in hydration *difference* experiments we expect the coherent scattering from the bulk of the deuterated protein to subtract out. However, two points need to be examined very critically. The first relates to the fact that in the low-hydration region the coherent scattering predominates; thus, for the small broadenings observed in this work, the possibility of residual contributions from line narrowing effects according to (7.1) and (7.2) arises (accepting the validity of the e.m.a. for present purposes). The other has to do with genuine changes in the coherent scattering as the result of a gradual 'loosening up' of the mantle of the protein molecule with increasing hydration. This is an effect of considerable interest in its own right. It is therefore important to examine the behaviour of the structure factor $S(k)$, which is proportional to the total intensity $I(k)$ measured in the powder diffraction experiment and shown in figure 6. Although it would perhaps be more appropriate here to use total intensity data derived directly from the corresponding IN 10 or IN 5 measurements, the D1B data are chosen in this semi-quantitative discussion because of their better angular resolution and uniform coverage of a large k region. The maxima in $\Delta E(k)$ according to figures 3 and 4 are characterized by values of k_{max} between 0.35 and 1.0 Å^{-1}, and this region corresponds to the $\theta = 4°$ to $12°$ interval in figure 6. The total intensity $I(k)$ increases very slightly in this region but is obviously devoid of any significant structure. The difference hydration levels, moreover, appear as essentially parallel tram-like traces narrowing almost imperceptibly (because of the Debye–Waller factor) with increasing k. We conclude that effects due to coherent scattering are unlikely to contribute more than perhaps 10% to the hydration difference broadenings observed in this k region. We can be less sure of this at higher k because of the broad feature in $I(k)$ centred on $2\theta = 16°$ or $k = 1.4$ Å^{-1}. The film-stack broadenings shown in figure 3 and the two highest hydration levels in figure 4 possess minima that partly overlap this feature, and it is possible that there is a larger coherent contribution here. The difference broadenings $\Delta E(k)$ obtained for the hydration sequence in figure 4 show both a systematic shift in their maxima from $k_{max} = 0.45$ to 0.7 Å^{-1} and increases in peak intensity with increased hydration (for a discussion of the hydration numbers, see §8 below). Although several aspects of the $\Delta E(k)$ data obtained so far need to be examined in greater detail by further experiments, the results considered as a whole suggest an interpretation in terms of water molecule motions between discrete sites separated by distances of the order of $1/k_{max}$. This broad conclusion will therefore be adopted as a working hypothesis in the remainder of this paper.

The migration of water molecules on or near the surface of a protein cannot be expected to conform to any kind of regular lattice of hydration sites. To extract some characteristic lengths and times from the data, we have nevertheless attempted to fit simple Chudley–Elliott (C.E.) jump diffusion models (Chudley & Elliott 1961) to the $\Delta E(k)$ data shown in figure 3. If it is assumed that a scattering centre (i.e. a water molecule) moves from its site at r to a new site at $r + a_n$ during $\Delta t \ll \tau_0$, and if there is a spatial distribution of n_s such sites accessible to this centre, then the scattering law for nearest-neighbour jumps may be calculated from a rate equation for the probabilities involved. The result is a Lorentzian of width

$$\Delta E(k) = \frac{2\pi}{n_s \tau_0} \sum_{n=1}^{n_s} [1 - \exp(-i\mathbf{k} \cdot \mathbf{a}_n)], \tag{7.3}$$

which has been evaluated explicitly or numerically for various geometrical arrangements of sites (Gissler & Rother 1970; Springer 1972). Because of the small number of detectors that can be accommodated in backscattering geometry during a single IN 10 experiment, the structure of $\Delta E(k)$ is only seen rather coarsely and any curve fitting is subject to this limitation. Thus the curves shown in figure 3 and the numbers derived from them must at present be regarded as illustrative of the technique rather than definitive. Except at very low k, the film-stack broadenings agree reasonably well with an isotropic jump model for which $\Delta E \sim 1 - \sin(ka)/ka$. For the uniform jump distance assumed here, we obtain $a = 5.5$ to 6 Å, and for the residence time $\tau_0 = 5$ to 10 ns. The broadenings measured for the 'fluffy' sample, on the other hand, exhibit a pronounced first minimum at $k_{min} = 0.8$ to 0.9 Å^{-1} that cannot be modelled in any simple way. The lower curve in figure 3 represents a weighted superposition of $\Delta E(k)$ resulting from one-dimensional and two-dimensional C.E. models, the reasoning for this being that at lower hydration levels one might expect the migration of water molecules to proceed preferentially along polypeptide chains at the surface. Here $a = 7$ to 9 Å and $\tau_0 = 15$ to 30 ns. This sample was hydrated continuously from the dry state at a very low rate; the hydration number at the end of the experiment, estimated from the increase in total neutron counts, was 300 ± 50 H$_2$O per subunit.

8. Hydration numbers and topography

Owing to limited instrumental accessibility, it was not possible in these first experiments to equilibrate the samples at different r.h. levels as well as we would have liked. It was necessary to adopt an operational rather than a thermodynamic definition of equilibrium, in the sense that the sample was considered equilibrated at a given humidity whenever the total count rate of scattered neutrons appeared to be reasonably close to its asymptotic value. The backscattering spectrometer was programmed to write a set of spectra accumulated during 30, 60 or 90 min into a memory, and the first 2–5 data blocks recorded between two successive r.h. levels were discarded upon inspection of the continuously monitored total count rate.

In the absence of direct gravimetric data, the number N_h of H$_2$O molecules per 29 000 molecular mass subunit of phycocyanin may be determined either by an absolute calibration (with respect to vanadium) of the increase in the total count rate of scattered neutrons relative to the count rate per unit mass of the dry sample, or by relating this increase to a calculated cross section for the protein molecule. The latter approach was chosen here because it proved to be necessary, with a view to a more detailed interpretation of succeeding experiments on deuteron-exchanged samples, to develop a model for the changes in cross section of a deuterated phycocyanin subunit upon H–D exchange. Approximate values for the coherent cross section at high k were obtained by splitting each of the 17 amino acid residues in two or three ways into fragments of 20–40 Å^3 (corresponding to $k \approx 2$ Å^{-1}), averaging over their coherent scattering lengths $b_{coh, i}$ and summing the cross-section averages $4\pi \langle b_{coh, i}\rangle^2$ for the residue composition given by Berns et al. (1964). This procedure may be regarded as a crude variant of the 'cube method' described by Federov et al. (1972). A set of 16 residues each containing two or three impurity hydrogens bound covalently to carbon were mixed in to simulate an isotopic impurity level of 98%. This gave $\sigma_{coh} = 100 \pm 15$ kb† for the proton-exchanged subunit; the corresponding incoherent cross section was $\sigma_{inc} \approx 40$ kb. The total cross section of 140 kb was then used to scale the intensity increases taken from the D1B data at $k = 2$ Å^{-1} (averaged over $\Delta k = 0.3$ Å^{-1}; compare figure 6) in terms of water uptake.

The unique advantage of neutron inelastic scattering to allow energy changes to be studied with spatial information deduced from their k dependence may be further exploited by linking the positions of the maxima in figure 4 with some model of the topography of hydration sites. While it is straightforward to obtain reasonably accurate values for the specific water uptake

† 1 barn (b) = 10^{-28} m^2.

of the sample as a whole and thus to calculate average hydration numbers, it is quite impossible to determine an average value for the area per molecule over which a process of surface diffusion would be effective. This would require a great amount of detailed and largely unknown information about the surface and internal structure of amorphous protein powders of the kind described in §4, the contact of multimeric units, etc. Accepting the interpretation of a surface jump diffusion process advanced in §7, it is possible, however, to argue that there must be a limiting value for the average distance between primary hydration sites that could be

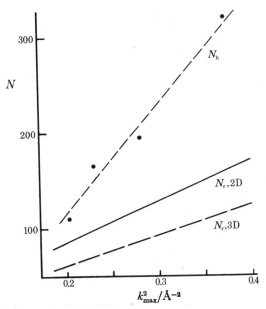

FIGURE 7. Hydration numbers N_h and site numbers N_r (for two-dimensional and three-dimensional Chudley–Elliott models) as a function of k_{max}^2 for the sample and hydration conditions of figure 4.

obtained from data of this kind at very low hydration levels. Because of the decreased scattering intensity at low k and low N_h, a direct measurement of k_{max} closer to its limiting value $k_{max,0}$ does not seem feasible at present with a backscattering spectrometer such as IN 10, although this should be possible with the neutron spin-echo spectrometer IN 11 at Grenoble (Richter et al. 1978). By extrapolating the curve connecting the maxima in figure 4 down to the abscissa, we obtain an approximate value for $k_{max,0}$ of 0.35–0.40 Å^{-1}. On the basis of simple C.E. models, this would convert into an average jump distance d_{r_0} of 10–12 Å, a quantity characterizing the distribution of primary hydration sites. To obtain an estimate of the number N_{r_0} of such sites, it is necessary to know the fraction of accessible surface area lost on association into an oligomer. This should be about 30–40 % (Teller 1976) and the result for N_{r_0}, using $d_{r_0} = 11$ Å, is 32–38. To illustrate these relationships we have, somewhat arbitrarily, scaled the water uptake corresponding to the first *measured* value of k_{max} such that each of these sites is occupied, on average, by three water molecules. Hence $N_h \approx 110$ at $k_{max} = 0.45$ Å^{-1}. By using values for the water uptake calculated from D1B data as described above, hydration numbers of 165, 195 and 310 were then determined for the sequence of $\Delta E(k)$ curves shown in figure 4. These N_h values are plotted separately in figure 7 as a function of k_{max}^2, together with two lines showing the dependence of N_r evaluated from (7.3) for a two-dimensional or three-dimensional C.E. model relating d_r and k_{max}. While the scaling of the reference value is uncertain on the basis of the present data, it is seen that the slope of N_h with respect to k^2_{max} differs significantly from that of either of the theoretical curves. This may be interpreted as an increased clustering of water molecules at hydration sites together with a gradual increase in the number of such sites, but may also in part be due to inadequacies in the simple jump diffusion model used. Data from H-D contrast experiments are likely to shed further light on this problem.

We are greatly indebted to all the following: D. S. Berns, Albany Medical College, New York, and H. L. Crespi, Argonne National Laboratory, Illinois, for deuterated C-phycocyanin; S. Gilmour, who took part in exploratory experiments, and A. D. Taylor, Rutherford and Appleton Laboratories, for data analysis on the Harwell results and subsequent participation at the I.L.L.; the Institut Laue–Langevin, Grenoble, for facilities; the I.L.L. staff scientists A. Heidemann, W. S. Howells, G. T. Jenkin (IN 10); A. J. Dianoux, F. Douchin, R. E. Lechner (IN 5); and P. Convert, G. Bomchil and J. L. Buevoz (D1B) for much advice. This work is supported by the Science Research Council.

REFERENCES (Middendorf & Randall)

Allen, G. & Higgins, J. S. 1973 *Rep. Prog. Phys.* **36**, 1073–1133.
Bacon, G. E. 1962 *Neutron Diffraction*, 2nd edn. Oxford: Clarendon Press.
Berendsen, H. J. C. 1975 In *Water, a comprehensive treatise* (ed. F. C. Franks), vol. 5, ch. 6, pp. 293–349. New York: Plenum Press.
Berns, D. S. 1971 In *Subunits in biological systems* (ed. S. N. Timasheff & G. D. Fasman), vol. 5A, pp. 105–148. New York: Dekker.
Birr, M., Heidemann, A. & Alefeld, B. 1971 *Nucl. Instrum. Meth.* **95**, 435–439.
Brunauer, S., Emmett, P. H. & Teller, E. 1938 *J. Am. chem. Soc.* **60**, 309–319.
Bull, H. B. 1944 *J. Am. chem. Soc.* **66**, 1499–1507.
Bull, H. B. & Breese, K. 1968 *Arch. Biochem. Biophys.* **128**, 488–496.
Chudley, C. T. & Elliott, R. J. 1961 *Proc. phys. Soc.* **77**, 353–361.
Clementi, E., Corongiu, G., Jönsson, B. & Romano, S. 1979 *FEBS Lett.* **100**, 313–317.
Clementi, E., Ranghino, G. & Scordamaglia, R. 1977 *Chem. Phys. Lett.* **49**, 218–224.
Cooper, A. 1976 *Proc. natn. Acad. Sci. U.S.A.* **73**, 2740–2741.
Convert, P. 1975 Thèse d'État, Université de Grenoble.
Crespi, H. L. 1977 In *Stable isotopes in the life sciences*, pp. 111–121. Vienna: I.A.E.A.
Dahlborg, U. & Rupprecht, A. 1971 *Biopolymers* **10**, 849–863.
Edsall, J. T. & McKenzie, H. A. 1978 *Adv. Biophys.* **10**, 137–207.
Egelstaff, P. A. 1967 *An introduction to the liquid state*. London: Academic Press.
Egelstaff, P. A. 1975 *Brookhaven Symp. Biol.* **27**, 1–26.
Egelstaff, P. A., Gray, C. G., Gubbins, K. E. & Mo, K. C. 1975 *J. statist. Phys.* **13**, 315–330.
Federov, B. A., Plitsyn, D. B. & Voronin, L. A. 1972 *FEBS Lett.* **28**, 188–190.
Finney, J. L. 1977 *Phil. Trans. R. Soc. Lond.* B **278**, 3–32.
de Gennes, P. G. 1959 *Physica* **25**, 825–839.
Gissler, W. & Rother, H. 1970 *Physica* **50**, 380–388.
Glauber, R. J. 1962 *Lectures in theoretical physics*, vol. 4. New York: Interscience.
Hagler, A. T. & Moult, J. J. 1978 *Nature, Lond.* **272**, 222–226.
Hermans, J. & Rahman, A. 1976 *CECAM Rep.* no. II.4, pp. 153–158.
Hopfinger, A. J. 1977 In *Intermolecular interactions and biomolecular organisation*, ch. 5, pp. 113–143. New York: Wiley Interscience.
Jacrot, B. 1976 *Rep. Prog. Phys.* **39**, 911–953.
Karplus, M. & Weaver, D. L. 1976 *Nature, Lond.* **260**, 404–406.
Kneale, G. G., Baldwin, J. P. & Bradbury, E. M. 1977 *Q. Rev. Biophys.* **10**, 485–517.
Kuntz Jr, I. D. & Kauzmann, W. 1974 *Adv. Protein Chem.* **28**, 239–345.
Larsson, K. 1971 *Phys. Rev.* A **3**, 1006–1022.
Lechner, R. E., Volino, F., Dianoux, A. J., Douchin, F., Hervet, H. & Stirling, G. C. 1973 *Multichopper IN 5 Scientific Status Report* (I.L.L. Report no. 73L85).
Lee, B. & Richards, F. 1971 *J. molec. Biol.* **55**, 379–400.
Levitt, M. 1976 *CECAM Rep.* no. III.6, pp. 219–228.
McCammon, J. A., Gelin, B. R. & Karplus, M. 1977 *Nature, Lond.* **267**, 585–590.

Maconnachie, A. & Richards, R. W. 1978 *Polymer* **19**, 739–762.
Mathur-de Vré, R. 1979 *Prog. Biophys. molec. Biol.* **35**, 103–134.
Middendorf, H. D. 1974 *Nucl. Instrum. Meth.* **114**, 397–399.
Nedev, K. N. & Churgin, Y. I. 1975 *Molek. Biol., Moscow* **9**, 761–767.
O'Carra, P. & O'hEocha, C. 1976 In *Chemistry and biochemistry of plant pigments* (ed. T. W. Goodwin), vol. 1, pp. 328–376. London: Academic Press.
Packer, K. J. 1976 *Phil. Trans. R. Soc. Lond.* B **278**, 59–87.
Randall, J. T., Middendorf, H. D., Crespi, H. L. & Taylor, A. D. 1978 *Nature, Lond.* **276**, 636–638.
Richter, D., Hayter, J. B., Mezei, F. & Ewen, B. 1978 *Phys. Rev. Lett.* **41**, 1484–1487.
Rowe, J. M., Sköld, K., Flotow, H. E. & Rush, J. J. 1971 *Phys. Chem. Solids* **32**, 41–54.
Rüdiger, W. 1980 In *Pigments in plants* (ed. F. C. Czygan). Stuttgart: Fischer Verlag. (In the press.)
Sköld, K. & Nelin, G. 1967 *J. Phys. Chem. Solids* **28**, 2369–2380.
Springer, T. 1972 *Springer tracts in modern physics* (ed. G. Höhler), vol. 64, pp. 1–100. Berlin: Springer-Verlag.
Steele, W. A. & Pecora, R. 1965 *J. chem. Phys.* **42**, 1863–1871.
Stuhrmann, H. B. & Miller, A. 1978 *J. appl. Cryst.* **11**, 325–345.
Taecker, R. G., Crespi, H. L., DaBoll, H. F. & Katz, J. J. 1971 *J. Biotechnol. Bioengng* **13**, 779–793.
Teller, D. C. 1976 *Nature, Lond.* **260**, 729–731.
Vineyard, G. H. 1958 *Phys. Rev.* **110**, 999–1010.
Volino, F. & Dianoux, A. J. 1978 In *Organic liquids: structures, dynamics and chemical properties*, pp. 17–47. Chichester: Wiley.
White, J. W. 1975 *Proc. R. Soc. Lond.* A **345**, 119–144.
Whittemore, W. L. 1968 In *Proc. 4th Symp. on Neutron Inelastic Scattering*, IAEA, Vienna, pp. 175–181.
Williams, R. J. P. 1978 *Proc. R. Soc. Lond.* B **200**, 353–389.
Winfield, D. J. & Ross, D. K. 1972 *Molec. Phys.* **24**, 753–772.
Worcester, D. L. 1976 In *Biological membranes* (ed. D. Chapman & D. F. H. Wallach), vol. 3, pp. 1–48. New York: Academic Press.

Discussion

A. J. LEADBETTER (*Department of Chemistry, University of Exeter, U.K.*). The quasi-elastic broadenings that the authors have observed are extremely small, being much less than the instrumental resolution width, and their conclusions about jump motion of water molecules are based on the Q-dependence of this very small broadening: $\Delta E(Q)$. Would the authors explain exactly what is the ratio between the incoherent, broadened quasi-elastic signal and the substrate background in their experiments?

Is not this substrate background scattering predominantly coherent with a maximum in $S(Q)$ near the position of the minimum observed in $\Delta E(Q)$? Can the authors therefore be quite sure that the structure observed in $\Delta E(Q)$ is not simply an artefact arising from the very small magnitude of ΔE and the high background scattering?

SIR JOHN RANDALL and H. D. MIDDENDORF. As to the magnitude of the effects seen at low levels of hydration, we feel that quasi-elastic broadenings of 0.8–4×10^{-3} cm^{-1} (f.w.h.m.), or 0.1–0.5 μeV, cannot be called 'extremely small' if observed in carefully designed and analysed difference experiments (compare, for example, Renouprez *et al.* 1979 (*J. chem. Soc. Faraday Trans.* I **75**, 2473) and Ewen & Richter 1978 (*J. chem. Phys.* **69**, 2954)). We wish to point out, in particular, that the broadenings shown were measured relative to the identical, completely undisturbed sample in successive runs, i.e. *not* relative to a vanadium plate, and that both methods of width analysis gave essentially the same results. We have shown previously (Randall *et al.* 1978, fig. 2) that the stability of IN 10 over uninterrupted run times of the order of 100 h is excellent for this purpose (r.m.s. intensity deviations less than about 2%). We regularly check this by concluding a hydration sequence with a long desorption run to get back to starting conditions. We see no reason why the potential of this and similar instruments for difference spectroscopy should not be exploited more fully in a way familiar to biophysicists and biochemists. The main limiting factors at present are the small number of spectra obtainable simultaneously and their non-uniformity as regards line width and shape; these technical drawbacks are being overcome by advanced instruments currently under construction (see contributions by Fender *et al.* and Springer, this symposium). In current neutron spectrosopic work on hydrogeneous biopolymers, the IN 10 line width changes seen on hydration of collagen are comparable to those that we report here, and the inelastic difference spectra observed on IN 5 for lysozyme and DNA are between 6 and 40 % of the 'background' spectrum (see *I.L.L. Annual Report* 1977, annex, pp. 374–377).

With regard to the role played by the coherent scattering, it was not possible because of the time limitation to discuss this adequately in our talk and details may be found in the text. It is important here, as you point out, to consider the variation of $\Delta E(k)$ in conjunction with $S(k)$ and therefore to collect accurate diffraction data in parallel with inelastic experiments. Owing to the difficult beam time situation, this was begun only recently and a full set of structure factor data have not been obtained. It is clear, however, from the limited results shown in figure 6 that the broad feature centred on $k = 1.4$ Å^{-1} neither coincides with the minimum of $\Delta E(k)$ nor shifts on hydration. While it may contribute to the depth and width of the dip in $\Delta E(k)$ at the highest hydration levels studied so far, all the available evidence points to the conclusion that line narrowing due to structure in $S(k)$ is not the dominant mechanism in determining the oscillatory behaviour of ΔE. We wish to emphasize that the results presented here are part of a continuing investigation into the hydration dynamics of *in vivo* deuterated biopolymers, and that phycocyanin is the first protein of this kind to be studied by quasi-elastic and inelastic scattering. Current H–D contrast experiments on 'fading out' the proton signal from the water of hydration and planned experiments on partly deuterated proteins will provide data enabling us to assess more quantitatively the contribution from coherent scattering.

The U.K. Spallation Neutron Source

By B. E. F. Fender[†], L. C. W. Hobbis[‡§] and G. Manning[‡]

[†] *Inorganic Chemistry Laboratory, South Parks Road, Oxford OX1 3QR, U.K.*
[‡] *Science Research Council, Rutherford Laboratory, Chilton, Didcot, Oxon. OX11 0QX, U.K.*

Pulsed neutron sources offer an attractive route for the realization of effective fluxes greater than those currently available from high flux reactors. The spallation neutron source now under construction at the Rutherford Laboratory will produce intense bursts of fast neutrons through interactions of 800 MeV protons with a heavy metal target. The fast neutrons are slowed down in nearby hydrogenous moderators viewed by some 20 time-of-flight neutron scattering instruments. The spectrum of the moderated neutrons is strongly enhanced in the high velocity region compared with that from a reactor. The new source will be comparable with the best beam reactors for experiments with neutrons of mid-thermal energy, and will provide unrivalled potential for use of the epithermal neutrons. Areas of science that will benefit immediately are the study of liquids and amorphous materials, high energy excitations in crystalline materials, molecular spectroscopy, surface phenomena and kinetic processes, as well as a range of crystallographic applications.

1. Introduction

A high intensity pulsed neutron source is being built at the Rutherford Laboratory for use in condensed matter research. The main attraction is that by using a high intensity proton accelerator to produce intense bursts of neutrons from heavy metal targets, effective neutron fluxes can be generated that, for many neutron scattering experiments, will be significantly greater than those available from the best steady-state reactor sources that exist today or are likely to be built in the foreseeable future. This paper aims to explain how this comes about and outlines the Spallation Neutron Source (S.N.S.) project (Hobbis *et al.* 1977) which started in mid-1977.

So far, most condensed matter research with neutrons has been done with the use of steady-state reactors, but even the most powerful reactor beams are very weak in comparison with the photon intensity from a laboratory X-ray source and ever higher neutron intensities are therefore continually in demand. However, it is generally agreed that the steady-state reactor is already close to its technological and financial limits. The use of pulsed neutron beams provides a solution.

2. Pulsed sources

In a diffraction experiment at a reactor, a monochromating crystal is normally used to define a narrow wavelength band of neutrons incident on the sample, and the various Bragg reflexions must be sought at different scattering angles; most of the neutrons in the initial beam are discarded. In the corresponding pulsed source experiment, time-of-flight is used to

[§] Present address: Science Research Council, Central Office, North Star Avenue, P.O. Box 18, Swindon SN2 1ET, U.K.

define the velocity (wavelength) of the diffracted neutrons and in principle none of the neutrons in the initial 'white' beam need be discarded; a fixed scattering angle can be used, although in practice several banks of counters are normally employed to improve counting rates. In the steady-state experiment a narrow wavelength band is used continuously; in the time-of-flight experiment a wide wavelength band is used for short periods.

For the ideal time-of-flight inelastic experiment, energy analysis by, for example, crystal reflexion measures final energy and is combined with overall time-of-flight data to define incident neutron energy in the so-called inverse geometry. Again a wide wavelength band pulsed incident beam is used. In contrast, in a steady-state triple-axis experiment a narrow wavelength band beam is incident on the sample. In inelastic experiments with choppers to define the incident energy, it is clearly the instantaneous pulsed source flux that has to be compared with the steady-state reactor flux.

TABLE 1. NEUTRON PRODUCTION PROCESSES AND SOURCES

process	neutron yield per incident particle	target energy per neutron MeV	practical sources		
			source	fast neutron yield/10^{15} ns^{-1}	target power MW
$(e\gamma)$, (γn) in heavy target	4×10^{-2}/60 MeV electron	1500	new Harwell linac	0.2	0.045
^{3}H $(^2$H n$)\alpha$	ca. 10^{-4}/300 keV deuteron	ca. 10^4	intense neutron source project, Los Alamos	ca. 1	0.3
fission	(ca. 1/fission)	ca. 200	I.L.L., Grenoble	ca. 2000	57
proton spallation non-fissile target	ca. 14/800 MeV proton	ca. 30	S.N.S., Rutherford Laboratory	ca. 20	0.09
fissile target	ca. 30/800 MeV proton	ca. 55		ca. 40	0.35

The general reasoning above, supported by detailed examination of representative experiments, leads to the conclusion that a pulsed source may achieve the same effective flux as a steady-state source but with greatly reduced mean power. Practical pulsed sources can be either pulsed reactors or accelerator based systems in which the burst of fast (megaelectronvolt) neutrons is first moderated in a small hydrogeneous moderator whose surface(s) form the effective source for the experiments. Ideally the pulse lengths of the moderated neutrons should not exceed a few tens of microseconds so that the required energy resolution can be obtained with reasonably short flight paths (10–100 m). Relevant experience with pulsed reactors exists in the Soviet Union, where a small (30 kW) system has been working since 1969 and a much larger system (4 MW, 10^{16} neutrons cm^{-2} s^{-1} peak intensity) is now in the final stages of construction. Unfortunately, pulsed reactors produce an inherently long pulse (the neutron bursts from the 4 MW reactor will be 90 μs long before moderation) and it would require in any case a very substantial development programme before such a source could be built in the U.K.

On the other hand, particle accelerators are well suited to the production of intense neutron bursts of ca. 10 μs duration and at 10–20 ms intervals, which are convenient for time-of-flight experiments. The most competitive accelerator sources are based on the use of protons to produce neutrons by spallation reactions from heavy nuclei; the use of a fissile material target yields additional neutrons from high-energy fissions. The high yield of neutrons from these reactions, combined with the low energy dissipated in the target per neutron produced, enables a neutron spallation source to have a performance comparable with or better than a reactor but with a heat dissipation in the target some 100 times lower than that of the reactor core. These features are illustrated in table 1, which includes data for neutron production by electron accelerators.

The potential of proton-induced spallation as a neutron source has long been recognized (Bartholomew & Tunnicliffe 1966), but it is only in recent years that accelerator technology has advanced far enough to permit the realization of a practical alternative to the steady-state reactor. A pulsed source with a proton linac is now in the early stages of use at Los Alamos Scientific Laboratory (Russell *et al.* 1978) and sources of intensity comparable with that of the Harwell Linac (Windsor *et al.* 1978) are being prepared at Argonne National Laboratory (Carpenter *et al.* 1978) and in Japan (Ishikawa & Watanabe 1978). Steady-state spallation sources can also have attractions, although they need relatively high average proton currents to be viable.

On a pulsed source it is essential to exploit the time structure of the neutron flux; so much so that the moderator is designed to optimize $I/\Delta t^2$ rather than I, where I is the time-averaged neutron flux and Δt is some measure of the pulse width (e.g. full width half maximum or standard deviation). A thin moderator made of a material that gives a high mean energy loss per collision is required to minimize Δt in the thermal region. Such moderators, in practice hydrogenous, greatly undermoderate the spectrum resulting in a $1/E$ slowing down component which, unlike that of a reactor, is of comparable intensity to the Maxwellian component. A pulsed source designed to have the same instantaneous flux in the thermal region as a particular reactor will have a performance some 1000 times better at 1 eV. The pulse structure in the slowing down region is sharp (e.g. $\Delta t = 2$ μs at 1 eV with $\Delta t \propto E^{-\frac{1}{2}}$) giving an intrinsically good time resolution. In the Maxwellian region, the pulse shape deteriorates with an attendant degradation of intrinsic resolution. Moderators of sub-ambient temperatures are used to extend the slowing down region to lower energies or neutron absorbers are introduced to suppress thermalization. Target–moderator coupling can be improved and the leakage of partly moderated neutrons reduced by surrounding the assembly with a neutron-reflecting material (Be, D_2O). A decoupling layer is required to prevent contamination of the moderator pulse by neutrons returning at long times from the reflector. A pulsed source target assembly is illustrated schematically in figure 1 while figures 2 and 3 illustrate some moderator characteristics. Specific choices in the design of the S.N.S. target–moderator–reflector system are given in §3e below.

It is clear that the spectral characteristics of pulsed sources favour the use of epithermal neutrons although, if the source is sufficiently powerful, thermal neutron experiments will not be excluded.

3. Description of S.N.S. facility

The source will use an 800 MeV high intensity proton synchrotron accelerator to deliver individual 400 ns bursts of 2.5×10^{13} protons at 50 Hz onto a heavy metal target to produce a fast neutron yield of *ca.* 4×10^{16} neutrons s^{-1}. Four moderators close to the target and embedded in reflector will slow these neutrons down to epithermal and thermal velocities and yield pulse lengths in the range 1–100 μs. The moderated neutrons will pass through 18 channels in a massive shield to the individual neutron scattering instruments ranged at distances of 6–100 m from the moderators. Table 2 gives the main parameters of the facility. Figure 4 illustrates its layout.

The choice of the main parameters (energy, proton intensity, repetition rate) was made on the basis of neutron yield from the moderators, overall costs and the need to use as much as possible of the former Nimrod plant and buildings. Neutron yields increase linearly with proton energy above 100 MeV and increase with increasing target mass number. Fissile target materials give significantly higher yields than non-fissile. Proton energies greater than about 1 GeV are not advantageous because of their longer range in targets and consequent extended source distribution; 800 MeV was a suitable energy from yield considerations and allowed the synchrotron to be fitted into the Nimrod magnet hall. Repetition rates above 50 Hz were ruled out by the economics of the laminated magnets and the r.f. accelerating system. The synchrotron magnet aperture, and hence the proton burst intensity, has been set as high as costs permit.

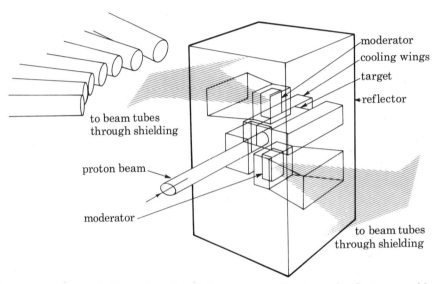

FIGURE 1. Schematic illustration of spallation target, moderator and reflector assembly, showing two moderators in wing geometry with single side viewing.

(a) Synchrotron main ring

The main synchrotron ring consists of a series of laminated bending and focusing magnets forming a ring of average radius 26 m. The proton beam circulates in a vacuum chamber maintained at no more than 5×10^{-7} Torr†. As the field in the bending and focusing magnets is increased, the beam is maintained centrally in the vacuum chamber by accelerating the protons with radio-frequency fields produced in r.f. cavities situated at six positions around the ring. The r.f. frequency, phase and amplitude are controlled to maintain the beam on a stable orbit as the field is increased from about 0.176 T at injection up to 0.7 T. This corresponds to the proton kinetic energy changing from 70 to 800 MeV as the circulation time changes from 1.49 to 0.65 µs. Bias windings on the ferrite in the r.f. cavities are used to sweep the accelerating frequency from 1.3 to 3.1 MHz. There are two equally spaced proton bunches circulating around the circumference of the accelerator ring.

† 1 Torr ≈ 133 Pa.

Tight focusing of the beam is needed to keep it within the 25 cm × 12 cm vacuum vessel aperture. The necessary transverse forces are provided by the quadrupoles, which control the transverse or betatron oscillations, and in the azimuthal direction the synchrotron oscillations are controlled by the r.f. accelerating field. Tolerances of the magnet parameters (fields and physical locations) must be carefully controlled to avoid unacceptable disturbance of the beam, and the betatron oscillation frequencies must be chosen to avoid the excitation of resonances in the motion by the residual errors. Magnetic correction elements are provided in the accelerator ring to reduce the effect of magnetic field errors. (Resonance excitation due to space-charge

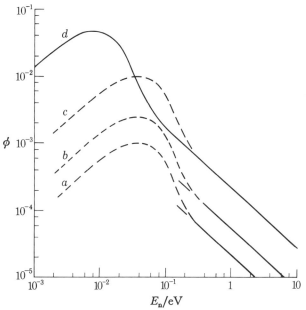

FIGURE 2. Computed neutron spectrum leaving moderators of 10 cm × 10 cm × 5 cm placed above a neutron source. The vertical scale is neutrons per steradian per electrovolt per fast neutron, and is for the total yield at 90° to the 10 cm × 10 cm face of the moderator. Curve (a) is for a bare water moderator at 300 K, 12 cm above a neutron source giving the spectrum expected from a small target (peak of the neutron spectrum at 1.4 MeV). Curve (b) is for the same moderator but with a neutron source characteristic of a large target (peak of the neutron spectrum at 0.14 MeV). Curve (c) is similar to (b) but with a reflector and decoupler. Curve (d) is similar to curve (c) but with a 10 cm × 10 cm × 5 cm moderator at 77 K. Experimentally measured spectra closely confirm the computations.

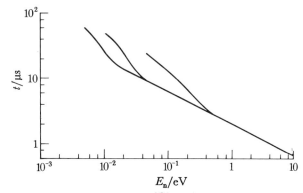

FIGURE 3. Full width at half height of the neutron time distribution leaving a 10 cm × 10 cm × 5 cm moderator as a function of neutron energy.

TABLE 2. S.N.S. MAIN PARAMETERS

injector

output energy	70.51 MeV
input energy	0.665 MeV
accelerated particle	H$^-$
pulse current	20 mA
pulse length	500 µs (max.)
r.f. frequency	202.5 MHz
injection method	foil stripper

main ring

number of superperiods	10
number of dipoles	10
number of doublet quadrupoles	20
number of trim quadrupoles	20
number of singlet quadrupoles	10
mean radius of synchrotron	26.0 m
repetition frequency	50 Hz
maximum proton energy	800 MeV
proton burst intensity	2.5×10^{13} per pulse

vacuum vessel

material	alumina ceramic
pressure	$< 5 \times 10^{-7}$ Torr

r.f. system

number of r.f. cavities	6
number of accelerating gaps/cavity	2
r.f. power (peak)	1.1 MW
r.f. frequency	1.34–3.09 MHz

extraction system

field in pulsed kickers (3)	0.040 T
kicker field rise time	0.225 µs
d.c. septum magnet field	1 T
proton burst length	0.4 µs

target station

target	Zircaloy-2 clad ^{238}U
target power	350 kW
fast neutron production rate (time average)	4×10^{13} s^{-1}
shield thickness nominal:	
forward direction	5 m
laterally	4 m
no. of beam channels	18

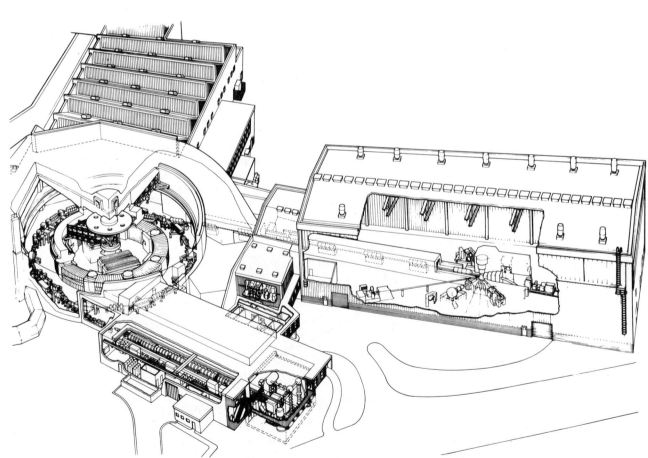

FIGURE 4. Illustration of the spallation neutron source that is being constructed at the Rutherford Laboratory. The linear accelerator used to inject the H$^-$ beam is at the bottom of the diagram, the synchrotron ring is at the left and the spallation neutron target station at the right. Only a few of the neutron scattering instruments are indicated. The buildings were formerly used for the Nimrod 7 GeV proton synchrotron and particle physics experiments.

effects cannot be so readily corrected.) The r.f. parameters must also be accurately controlled, with fast feedback loops playing an important part.

The magnets in the ring are tuned by a large capacitor bank and driven at approximately 50 Hz by a biased sinusoidal voltage. The d.c. component sets the minimum field corresponding to injection at 70 MeV. Circuit losses will be made up from a motor alternator set whose frequency must be carefully chosen to be compatible with the use of choppers in the neutron beams. A choice will be made between operation at mains frequency, at the natural resonant frequency of the magnet, or by locking onto a fixed frequency determined by a crystal oscillator, which also controls the choppers.

Inside the magnets, the vacuum vessels must be electrically insulating to avoid eddy current heating and will be made of high alumina ceramic chosen for its good vacuum properties and high thermal conductivity. Elsewhere, stainless steel chambers will be used. A pressure of less than 5×10^{-7} Torr is required to avoid beam/gas instabilities and will be realized with 30 triode ion pumps. An r.f. shield (Faraday cage) must also be provided inside the vessel to ensure that beam instabilities are not produced by electromagnetic coupling to its surroundings.

Non-destructive beam monitors are required to measure the beam intensity and position in space and time. They form an essential part of the beam control servo-circuits as well as giving diagnostic information needed during commissioning and fault finding.

(b) Injection

The former 70 MeV Nimrod injector (an Alvarez, axial field r.f. linac with drift tubes) will be used, uprated for 50 Hz operation, and will provide a pulse current of 20 mA (West 1979). Multi-turn injection (315 turns during 470 µs) is needed to achieve the required synchrotron intensity. To avoid a phase space limitation on the number of turns that can usefully be injected, negative hydrogen ions will be delivered by the linac and be stripped by a foil (thickness *ca.* 50 µg cm^{-2}) to form H$^+$ as they enter the ring. About 50% of the beam will be trapped by the synchrotron r.f. system.

(c) Extraction and beam transport

The 800 MeV beam is extracted from the synchrotron by means of three full aperture ferrite kicker magnets with a rise time of *ca.* 200 ns followed by a d.c. septum magnet which together deflect the beam in the vertical plane into an external transport channel. The kicker field rises within the interbunch interval of *ca.* 250 ns, thus ensuring that the circulating beam is extracted within a single turn and a very short proton burst (*ca.* 450 ns) is delivered to the target. After extraction, the beam is transported by a channel containing 64 separate magnetic components over the distance of *ca.* 125 m to the target station. The extraction and transport system overall is achromatic and provides a 70 mm × 70 mm stable spot size at the target.

(d) Intensity limitations

In addition to any beam losses caused by failure to operate within the specified machine tolerances, there are many interactions that can cause the beam to become unstable as the beam current is increased (broadly speaking, space charge effects). The more important instabilities are:

(i) transverse incoherent space charge effects – a defocusing effect due to the mutual repulsion of protons within a bunch;

(ii) transverse coherent effects – the beam produces an image current in nearby components and the whole bunch is excited into coherent motion by the antidamping effect of the fields between the image currents and the bunch;

(iii) longitudinal instabilities – these can arise when a bunch induces fields in nearby cavity-like objects, which then interact with the same or subsequent bunches.

In the S.N.S., the use of the vacuum vessel r.f. shield and care in component design will contain the effects of (ii) and (iii), leaving the space charge effect, (i), to determine the intrinsic intensity limit at a level calculated to be 2.5×10^{13} protons per pulse.

Some beam losses are inevitable during the processes of injection, r.f. trapping, acceleration and extraction, but they must be strictly limited because of the radioactivity and radiation damage induced in the machine and other nearby components. A special collector system is being designed to localize the low-energy losses, and, generally, a combination of component design, local shielding and special handling equipment is needed to ensure that an acceptable operating schedule can be maintained. Even with the advantage of a modern computer-based control system, commissioning up to routine operation at design level may take up to 2 years, about twice as long as for most accelerators. Neutron scattering experiments will, however, be possible throughout commissioning except for the very early stages.

(e) Target station

The fast neutron production target consists of *ca.* 30 plates of Zircaloy clad uranium-238 separated by cooling channels and forming an assembly 10 cm × 10 cm in cross section and 30 cm long with cooling manifolds on two sides. Neutron yields, energy deposition, residual activity and the nuclide inventory in a used target, have been computed by using codes developed from a well known Oak Ridge package, HETC (Chandler & Armstrong 1972). There is a significant contribution to the neutron production from high-energy fissions and from the target material outside the beam cross section. This target, dissipating 350 kW, will yield approximately 26 fast neutrons per incident proton, somewhat less than the frequently quoted 30 for solid uranium because of the diluting effects of the cooling channels and Zircaloy.

Of the fast neutrons, 0.6 % have energies exceeding 100 MeV and are emitted within a 15° semiangle forward cone; the average energy of all neutrons is less than 1 MeV. These yield predictions are used as the design basis of the bulk shield, which must provide an attenuation of approximately 10^9. There will be 5.2 m of iron and concrete in the forward direction, 4.2 m radially and 3.9 m above the target. The shield will rest on a steel plinth 2 m deep, made from former Nimrod magnet sectors.

Radiation damage growth effects are expected to limit the useful lifetime of a uranium target to not less than 3 months. Induced activity after this time will be approximately 400 kCi, comparable with a DIDO type fuel element, so a remote handling cell is provided immediately downstream of the target station to replace the used targets. Alternative targets, e.g. of tantalum, will be available should they be required.

The moderator geometry adopted is a four-wing arrangement with two moderators above and two below a horizontal target. (Wing geometry is analogous to a tangential beam on a reactor.) The front moderators are in the region of highest neutron flux and each is viewed on both faces. The rear moderators are mounted parallel to the target and are viewed on only one face. By suitable angling of the front moderators to the axis of symmetry of the target, an 18 beam channel layout may be achieved. The radiation and background levels experienced will be lower than those from slab moderators (analogous to a radial beam on a reactor), where a direct view of the target through the moderator is possible. To meet the requirements of the proposed instruments, moderators with a wide spectral range are required. In the slowing down region, optimization requires the moderator to have a source area of *ca.* 100 cm^2 and a thickness of *ca.* 5 cm, and the moderator material to have as high a hydrogen density as possible. Radiation damage effects and the need to remove up to several hundred watts of induced heat limit the choice to H_2O, liquid H_2 and some metal hydrides. In the Maxwellian

TABLE 3. S.N.S. MODERATORS

moderator location		number of beam channels	material	$\phi(E)_{1\,eV}$
upper front	A	6	ambient H_2O	2.2×10^{-4}
lower front	B	7	heterogeneously poisoned H_2O *or* cooled metal hydride	1.8×10^{-4}
upper rear	D	2	ambient H_2O *or* heterogeneously poisoned H_2O	1.3×10^{-4}
lower rear	C	3	liquid H_2	1.1×10^{-4}

$\phi(E)_{1\,eV}$ is the yield of neutrons at 1 eV per electronvolt per steradian per source neutron for a 10 cm × 10 cm × 5 cm moderator with the hydrogen density of H_2O. For comparison, the S.N.S. single moderator reference design value for $\phi(E)_{1\,eV}$ is 2.5×10^{-4}.

region, neutrons approach thermal equilibrium with the moderator material and a significant fraction of the pulse is contained in a long exponential tail. The resultant degradation of performance is avoided by lowering the physical temperature of the moderator or poisoning the moderator with a resonant absorber (Cd or Gd). A 20 K liquid hydrogen moderator of some 9 cm thickness is required specifically for the production of high intensities of long wavelength neutrons ($\lambda > 4$ Å†) for applications that do not demand the shortest possible pulse lengths.

The performance of the four-wing moderator system discussed above, as predicted by using programs based on the TIMOC Monte-Carlo code (Kschwendt & Rief 1970), is given in table 3. Calculations and measured results on benchmark systems (Boland *et al.* 1978) indicate an absolute accuracy of 20% and relative accuracy of 5%. The performances quoted are with a reflector of D_2O-cooled Be and decoupler energy 60 eV. (It is expected that the decoupler energy may be lowered somewhat, particularly for thermal instruments.) The flux of neutrons $n(E)$ having energies between E and $E + \Delta E$, on a sample at distance L is

$$n(E) = \phi(E)\,\Delta E n_f/L^2 \text{ neutrons cm}^{-2}\text{ s}^{-1},$$

where n_f is the fast neutron production rate in the target.

Comparison of the performance of pulsed and reactor sources is not straightforward. The direct comparison of the equivalent 4π flux emitted during the pulse and the steady 4π flux in the reactor shows that the intrinsic brightness of the S.N.S. is comparable with that of the Institut Laue–Langevin (I.L.L.) reactor in the thermal region, somewhat inferior in the low energy region served by the I.L.L. cold source, but is 10 times greater at 100 meV and 1000 times greater at 1 eV. Although intrinsic brightness is certainly not the sole factor in evaluating a source for a particular scientific experiment, it does in many cases provide a rough guide.

† 1 Å = 10^{-10} m = 10^{-1} nm.

4. Instruments and techniques

(a) Planned instruments

It is expected that the S.N.S. target station will eventually be able to support 20–25 instruments on 18 beam channels. For initial operation, it is planned to install 10–12 instruments with an overlapping programme of further instruments to follow. Seven have already been chosen for construction to begin now. Brief descriptions are listed in table 4. Five of these initial instruments will exploit the special advantages of the S.N.S. in the epithermal part of the neutron spectrum. Two instruments, the low Q spectrometer and the high resolution quasi-elastic spectrometer, are biased towards the use of long wavelengths from a cold (20 K) moderator where the effective fluxes are not so high as those at I.L.L. Nevertheless, these two instruments incorporate design features to give performances comparable with the equivalent instruments at I.L.L. (the small angle scattering apparatus (D 11 and the inelastic spectrometer IN 10).

(b) Envisaged instruments

Instruments following those listed in table 4 will incorporate new developments, including experience from the Harwell Linac and the S.N.S. itself. Some beam channels will be retained for completely novel instruments when they emerge. Additional instruments already proposed include:

single crystal diffractometer;	constant Q spectrometer;
small-angle neutron diffractometer for liquids and amorphous materials;	high symmetry spectrometer; eV spectrometer;
medium energy inelastic spectrometer;	polarization spectrometer.

Of the additional instruments listed above, a second liquids and amorphous materials machine is at the planning stage; this will be a small-angle instrument for measurements in the low Q region ($Q \approx 0.5$ Å^{-1}). Four instruments await experience at the Harwell Linac before their detailed design can be finalized. Tests will be conducted on the Linac of a simple single crystal diffractometer, of elastic polarization analysis with the use of two samarium filters (Freeman & Williams 1978), and of the use of resonance detectors (Ta, U) for measuring very high energy (0.5–5 eV) transfers such as would occur in studies of high energy excitations in crystalline materials. A constant-Q spectrometer, a time-of-flight equivalent of the triple axis machine (Windsor et al. 1978), is being installed on the Linac for use in the U.K. research programme (figure 6); construction of the S.N.S. instrument will benefit directly from this experience.

(c) Techniques

A number of important developments are incorporated in the S.N.S. instrument designs, notably in the fields of computing, detectors and neutron choppers as outlined below. In addition, a wide range of sample environment equipment (cryostats, furnaces, magnets, pressure cells, etc.) will be provided, standardized as far as possible between instruments, with computer control of variable parameters available. The high data rates and the inherently pulsed nature of the S.N.S. will be especially suited to the use of pulsed, or time-varying, environment parameters.

The instruments will depend crucially on the use of computers to control experiments, to acquire and organize data, and to convert it into a state where users can analyse it. Many of

TABLE 4. INITIAL BATCH OF S.N.S. INSTRUMENTS

instrument	outline description
high intensity powder diffractometer	An instrument of 'conventional' design to measure powder diffraction profiles rapidly and with higher statistical accuracy than currently possible while retaining a reasonable resolution in the lattice parameter ($\Delta d/d \approx 6 \times 10^{-3}$). See figure 5.
high resolution powder diffractometer	To complement the high intensity diffractometer when high spatial resolution is needed. High resolutions ($\Delta d/d \approx 3 \times 10^{-4}$) are achieved by means of a 100 m long neutron guide.
liquids and amorphous materials diffractometer	An instrument of basically conventional design deriving from the series used successfully on the Harwell Linac. Q range 0.2–60 Å^{-1}.
low Q (small-angle scattering) spectrometer	A 'small angle scattering' instrument in demand for a wide range of science. Improvements will be realized on existing reactor instruments by means of a sophisticated detector system. Q range 0.0015–4 Å^{-1}, continuously accessible.
high resolution quasi-elastic spectrometer	An instrument exploiting the time-of-flight method to advantage while using back scattering from analyser crystals to give very high resolution: $ca.$ 1 µeV within a 250 µeV window at final energy 2.07 meV (graphite analyser) or $ca.$ 13 µeV within 960 µeV at 1.82 meV (silicon analyser).
high throughput inelastic spectrometer	The time-of-flight version of the well known beryllium filter machine used for medium to high energy transfer work (20–500 meV) when high energy resolution and Q definition are not needed, especially useful for chemical spectroscopy with high sample throughout.
high energy inelastic spectrometer	An instrument using the latest developments in chopper technology to achieve high resolution at high energy. For example with the 3° forward detectors and incident energy 1 eV, an energy resolution of $ca.$ 4% is obtainable at final energy 0.4 eV and $Q \approx 5$ Å^{-1}.

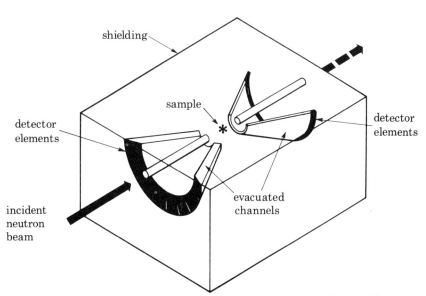

FIGURE 5. Sketch of the high-intensity powder diffractometer. The sample (indicated by the asterisk is viewed by two semicircular Li-glass scintillator detector banks situated at 30° and 150° to the incoming neutron beam and 1 m from the sample. The 150° detector is inclined in time-focusing geometry. The wavelengths of the diffracted neutrons are measured by time-of-flight. Resolution, $3 \times 10^{-3} < \Delta d/d < 10^{-2}$.

the problems arising from the high data rates and the use of the time-of-flight mode could not be solved economically without the recent and continuing rapid developments in the computing field. Software development will also need to break much new ground for the treatment of time-of-flight data; work has begun on profile refinement of time-of-flight powder diffraction patterns.

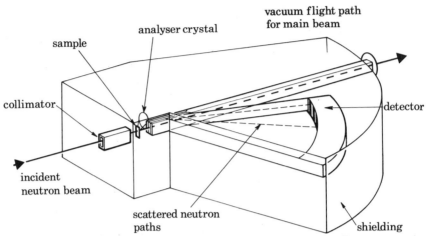

FIGURE 6. Sketch of the constant Q spectrometer planned for use on the S.N.S. in the field of lattice dynamics. The energies of inelastically scattered neutrons are measured by the analyser crystal and the detector array. The initial energies can then be calculated from the overall times of flight. The energy analyser is a large single crystal of germanium in transmission geometry and the detector bank uses 50 Li-glass scintillators, each 8 mm × 60 mm, mounted 1 m from the analyser.

Position sensitive detectors will be needed that are capable of detecting epithermal neutrons with good efficiency (up to 40 %), good gamma discrimination and short dead time (ca. 100 ns). Scintillation detector systems have been developed at the Rutherford Laboratory to meet these criteria. In a few cases, traditional gas counters will be used. The scintillator system modules have been tested in various experiments, including some at a pulsed spallation source, and will be incorporated for full-scale tests in some of the Linac instruments.

Several types of mechanical chopper are needed: a Fermi fast chopper with a burst time ca. 1 µs; fast neutron eliminators to reduce backgrounds; various disk choppers for selecting energy windows, time frames etc.; and mechanical velocity selectors for defining a variable wavelength window with resolution $\Delta\lambda/\lambda \approx 0.05$ to 0.5. The most demanding requirement is that of the Fermi fast chopper for the high energy inelastic spectrometer, where an energy resolution of the order of 1 % will be required at neutron energies up to 1 eV. A suitable slit package has been developed for this application by using aluminium–boron fibre composite in a rotor that can be driven by a standard Harwell spinning head (Jones et al. 1979).

5. Scientific applications

(a) Introduction

The S.N.S. will have an impact scientifically, just because for many experiments the effective neutron flux is increased relative to the best reactor sources. However, the way that the extra intensity is utilized will vary greatly from experiment to experiment. In some cases the researcher will wish for an increased data collection rate not only simply to improve the statistics but also, for example, to follow time-dependent phenomena; in other instances it will be the ability to measure weak scatterers that will be important. On several standard instruments such as the liquids and amorphous materials diffractometer, the choice of high count rates or the deployment of small or dilute samples remains open to the experimenter. In other instruments the availability of higher flux will be used to give improved resolution (e.g. in the high resolution powder diffractometer), or by using the epithermal bias of neutrons from a pulsed source, an extended range of energy or momentum transfers (e.g. in the high energy inelastic chopper spectrometer). Also quite apart from the emphasis on shorter wavelength neutrons, the use of a pulsed white beam is valuable for the simultaneous investigation of a wider range of energy and momentum transfer than can normally be obtained by using monochromated beams. Table 5 gives an indication of some of the areas of science that will benefit from these developments, which are further discussed below.

TABLE 5. THE IMPACT OF THE S.N.S. ON NEUTRON STUDIES

improvements envisaged	scientific areas affected
higher counting rates	(a) structure of fluids: more accurate partial radial distribution functions; the influence of temperature, pressure and concentration more easily studied; extension of isotopic substitution techniques (b) time-dependent structural changes readily followed (c) 'routine' molecular spectroscopy in the range 10–500 meV available
weak scatterers more accessible	(a) structural and dynamical studies of physisorbed or chemisorbed atoms and molecules (b) examination of dilute systems such as gases or matrix isolated molecules (c) general use of small samples where necessary, e.g. when rare isotopes are involved and amorphous thin films
better resolution in $Q, \hbar\omega$	(a) very high resolution powder diffraction, particle broadening effects (b) improved inelastic studies including separation of diffusional broadening and dispersion effects in the spectroscopy of molecular fluids and the minimization of multiphonon effects on solids (c) more powerful polarization analysis experiments for the separation of nuclear and magnetic scattering, coherent and incoherent scattering, magons and phonons; as well as the identification of magnetovibrational effects
more short λ neutrons; large $\hbar\omega$ (and/or Q)	(a) magnon studies up to 0.3 eV (b) improved chemical spectroscopy at high energies (c) spectroscopy in the electronvolt region (d) enhanced interatomic resolution in molecular fluids and amorphous solids (e) extended use of anomalous scattering techniques (f) studies possible on highly (neutron) absorbing materials
wide $Q/\hbar\omega$ window	(a) advantageous for certain studies of diffusion in molecular crystals and liquid crystals, polymers, intercalated compounds and biological membranes and for low-lying inelastic processes (b) more complete small-angle scattering data for studies in polymeric science, biology, metallurgy and the investigation of nucleation and growth processes

(b) *Liquids and amorphous solids*

In some simple systems, e.g. alkali halides, where the interactions can be described by pairwise potentials, the neutron results confirm the merits of the computer simulation studies. However, there is much to learn for most liquids, where chemical bonding is important. Similarly, Enderby's beautiful work on the structure of aqueous solutions is only a beginning

(Enderby, this symposium). Subtle details of the structure, including the changes with concentration, pressure and temperature, can only be revealed by data of very high statistical accuracy; hence the need for high rates of data collection. Because of the limited fluxes, the systems studied so far involve elements with readily available isotopes that have relatively large differences in scattering lengths; higher fluxes will broaden the range of systems to include those with rarer isotopes and smaller scattering length differences.

An important consequence of the high epithermal flux from the S.N.S. is the opportunity that it provides to extend upwards the range of momentum transfer and so to enhance the resolution in real space. Again, there have been some excellent pioneering studies on molecular liquids principally by the group from the University of Kent (see, for example, Walford *et al.* 1978). Improved real space resolutions will also aid the determination of the structure of amorphous solids (which also benefit from the use of isotopic substitution techniques). Higher fluxes will open the way to the study of much smaller samples (e.g. thin films) and an investigation of the links between structure and preparative conditions as well as kinetic investigations of the structural rearrangements that occur before phase separation or crystallization.

There will also be increased interest in the dynamics of fluids, whether it be in the study of diffusion processes, with the use of the high-resolution quasi-elastic spectrometer, or in the examination of vibrational modes, with the use of the high-energy inelastic spectrometer (at low Q). The latter instrument will also be suitable (at very high Q) for experiments on quantum liquids, where measurements of the proportion of ^{4}He atoms in liquid ^{4}He with the zero momentum state can be envisaged.

(c) Crystallography

The new source will provide outstanding facilities for neutron powder diffraction. On the high intensity instrument it would be possible to collect a complete set of good quality data in a moderately complicated structure (e.g. Al_2O_3) in about 30 s. As a result, a major use of the instrument is likely to be the use of variations in ambient conditions, e.g. electrical field and mechanical stress, to follow relaxation effects structurally. The high intensity available will also greatly improve the prospects for accurately detecting very weak features arising, for example, from atoms or molecules adsorbed on a surface, so that it is possible to envisage present studies (Bomchil *et al.*, this symposium) being extended to the investigation of the build-up of chemisorbed layers on relatively low surface area metals.

The high resolution powder diffractometer is designed to extend to the full the power of profile analysis methods, which have proved so successful in recent years (Cheetham & Taylor 1977) and which are now available for time-of-flight instruments (R. B. Von Dreele, personal communication, 1979). An improvement in resolution up to $\Delta d/d$ of 3×10^{-4} will both make it much easier to determine unknown structures and increase the number of independent structural parameters that can be refined to perhaps as high as 200. This instrument and the high-intensity diffractometer will be equipped with ring detectors, which can detect anisotropy in the radial scattering, thus providing excellent opportunities for textural studies of which the recent stress effects in bone provide a very good example (Bacon *et al.* 1979).

The success of time-of-flight single-crystal diffraction will depend on the development of area detectors and on solving the problems of data collection. However, given that wavelength-dependent structure factors would be routinely available for extinction corrections, very accurate measurements are in principle possible. In these circumstances a single crystal diffractometer on the S.N.S. would be particularly powerful for structure determination involving unit cells up to 20 Å and for the determination of very high resolution data.

(d) Molecular spectroscopy

The ubiquity of hydrogen and its large incoherent cross section for neutrons means that there is potentially a heavy demand for neutron (proton) spectroscopy. Bands that are weak or forbidden (for symmetry reasons) in the infrared or Raman regions may be seen by neutrons, and optically opaque materials such as metals are often transparent to neutrons.

At present, neutron spectroscopy is curtailed by a limited energy range, relatively low count rates and poor resolution. The S.N.S. high throughput inelastic spectrometer will enable normal spectroscopic measurements in the energy range $10\,\mathrm{meV} < \hbar\omega < 500\,\mathrm{meV}$ to be made in minutes rather than hours, although still with modest resolution. It is therefore possible to envisage experiments involving a series of samples that parallel those employing standard laboratory infrared instruments. The possibility for improving resolution at the expense of counting times will exist and, of course, the high intensity will be particularly valuable in the examination of hydrogenous molecules (or atoms) in dilute solution or as an absorbed species.

(e) Polymer science and biology

One of the recent successes of neutron scattering has been the very rapid growth in experiments on polymers and biological materials. This has been associated to a large extent with instruments at the I.L.L. that make use of long wavelength neutrons. Although the flux of cold neutrons is more modest from the S.N.S. it is nevertheless clear that a powerful low Q instrument can be built that is capable of sustaining programmes similar to those described by Jacrot (this symposium). Because the S.N.S. instrument employs a wide band of wavelengths, it will cover, in one experiment, a much larger Q range than D 11 at the I.L.L. This is important when the dimensions of the scattering particles are deduced from an analysis based on both Guinier and Porod data together with the evaluation of an integrated intensity. For quasi-elastic and low-energy inelastic measurements the high resolution quasi-elastic scattering spectrometer will also, because of its versatility, prove to be competitive with I.L.L. instruments.

(f) Solid state physics

Many of the neutron experiments currently carried out involve magnon and phonon studies on triple-axis spectrometers. In §4 the equivalent time-of-flight spectrometers are mentioned; these instruments, which will allow larger energy transfers to be studied, will have many applications. One of the most exciting features of the S.N.S. however, is the prospect that the effective fluxes will be high enough to carry out useful polarization analysis experiments. In magnetic inelastic scattering the following areas of interest can readily be identified in (i) the separation of coherent magnetic excitations from other scattering (for example, in cases where the magnon energies overlap considerably with high energy optic phonon branches), (ii) the measurement of single particle excitations in ordered magnets (of considerable significance to the theory of magnetism in transition metals) and (iii) the measurement of inelastic paramagnetic scattering in both ordered magnets above the transition temperature and in dilute alloys. Again, the results of such experiments are likely to challenge directly our theoretical understanding of magnetism at the atomic level. Polarization analysis also allows the separation of spin incoherent scattering from coherent scattering so that for some elements, especially hydrogen, nuclear inelastic scattering gives information about the self and pair dynamic correlations. The inelastic polarization analysis experiments, in terms of instrument design, will be among the most demanding on the S.N.S.

Somewhat less demanding, but still important, are total scattering polarization analysis measurements, which allow an identification of the elastic magnetic scattering. For paramagnets and spin glasses and for amorphous ferromagnets with large magnetic anisotropy, it is very difficult to separate the elastic magnetic component of the scattering without polarization analysis.

Spectroscopy in the electronvolt region is a unique possibility where very high energy transfers (0.5–5 eV) are needed for investigating high-energy excitations in crystalline materials, including interband transitions in semiconductors, and may also be useful for the study of single particle distribution functions, e.g. in liquid helium.

This paper is based on the work of the S.N.S. project team at the Rutherford Laboratory and members of the neutron scattering community participating in the instrument design, to all of whom acknowledgement is made.

References (Fender *et al.*)

Bacon, G. E., Bacon, P. J. & Griffiths, R. K. 1979 The orientation of apatite crystals in bone. *J. appl. Crystallogr.* **12**, 99–103.

Bartholomew, G. A. & Tunnicliffe, P. R. (eds) 1966 The AECL study for an intense neutron-generator. *Atomic Energy of Canada Ltd report* no. AECL-2600.

Boland, B. C., Carne, A., Stirling, G. C. & Taylor, A. D. 1978 Spallation target-moderator-reflector studies on Nimrod. *Rutherford Laboratory report* no. RL-78-070.

Carpenter, J. M., Price, D. L. & Swanson, N. J. (eds) 1978 IPNS – a national facility for condensed matter research. *Argonne National Laboratory report* no. ANL-78-88.

Chandler, K. C. & Armstrong, T. W. 1972 Operating instructions for the high-energy nucleon–meson transport code HETC. *Oak Ridge National Laboratory report* no. ORNL-4744.

Cheetham, A. K. & Taylor, J. C. 1977 Profile analysis of powder neutron diffraction data: its scope, limitations, and applications in solid state chemistry. *J. Solid State Chem.* **21**, 253–275.

Freeman, F. F. & Williams, W. G. 1978 A ^{149}Sm filter for thermal neutrons. *J. Phys.* E **11**, 459–467.

Hobbis, L. C. W., Rees, G. H. & Stirling, G. C. (eds) 1977 A pulsed neutron facility for condensed matter research. *Rutherford Laboratory report* no. RL-77-064/C.

Ishikawa, Y. & Watanabe, N. 1978 KEK neutron source and neutron scattering research facility. *National Laboratory for High Energy Physics (Japan) report* no. KEK-78-19.

Jones, T. J. L., Penfold, J. & Williams, W. G. 1979 The use of boron fibre/aluminium laminates in neutron beam instrumentation. *Rutherford Laboratory report* no. RL-79-020.

Kschwendt, H. & Rief, H. 1970 TIMOC – a general purpose Monte-Carlo code for stationary and time-dependent neutron transport. *Euratom report* no. EUR 4519e.

Russell, G. J., Lisowski, P. W. & King, N. S. P. 1978 The WNR facility – a pulsed spallation neutron source at the Los Alamos Scientific Laboratory. In *Proceedings of the International Conference on Neutron Physics and Nuclear Data for Reactors and other Applied Purposes*, Harwell, September 1978, pp. 1135–1140. OECD/NEA.

Walford, G., Clarke, J. H. & Dore, J. C. 1978 Neutron diffraction studies of liquid carbon suboxide. I. Molecular conformation and orientation correlation. *Molec. Phys.* **36**, 1581–1600.

West, N. D. 1979 The Rutherford Laboratory 70 MeV linear accelerator injector. In *Proceedings of the 1979 Proton Linear Accelerator Conference*, Brookhaven National Laboratory. (In the press.)

Windsor, C. G., Heenan, R. K., Boland, B. C. & Mildner, D. F. R. 1978 A constant Q spectrometer for pulsed neutron sources. *Nucl. Instrum. Methods* **151**, 77–488.

Developments in experimental neutron physics at the Institut Laue–Langevin

By T. Springer

Institut Max von Laue–Paul Langevin, Boîte postale, Centre de tri 156 X, F-38042 Grenoble-cedex, France

A report and survey are given dealing with a variety of recent developments in neutron spectroscopy, diffraction and beam techniques, referring to work at the Institut Laue–Langevin. Subjects such as high resolution spectroscopy by 90° Bragg diffraction and spin echo analysis (leading to 10^{-7} to 10^{-8} eV energy resolution), spin polarizers, focusing crystals, modern multicounters, and methods dealing with neutrons of energies in the 10^{-7} eV region (ultra-cold neutrons), are dealt with.

Introduction

Besides the operation of the high flux reactor for the scientific users, a traditional and genuine task of the Institut Laue–Langevin (I.L.L.) is the innovation, development and improvement of the instruments in neutron physics. Regarding in particular diffractometry and spectroscopy, this task aims essentially at two aspects: first, the improvement of the statistical accuracy of the data and, secondly, the refinement of the instrumental resolution in energy and momentum space, both aspects being naturally related in most cases. This article presents a short review of recent methodical developments connected with these subjects, namely

high resolution spectroscopy;

methods of enhancing intensities and/or count rates, by increasing solid angles, and by multi-counters; and

production of high-intensity polarized neutrons.

The review also deals with the methods of achieving high fluxes or densities of ultra-cold neutrons.

Other means to enhance the accuracy of the results, namely speeding up the instrument control by computers and by sophisticated interaction between results and instrument operation, will not be treated. Nevertheless, we consider these aspects to be as important as those mentioned before. Many of the developments to be treated, e.g. multicounters, are so efficient that they also allow a low flux reactor to serve as a tool for many interesting applications.

High resolution spectroscopy

At present, we envisage four fundamental ways to define energies and energy transfers in neutron spectroscopy: (i) Bragg reflexion; (ii) measurement of flight-times with fast chopping devices; (iii) time-of-flight by using the 'inherent clock' of a neutron passing through a magnetic field and finally (iv) comparing the neutron's kinetic energy with interacting fields, such as gravity (Steyerl 1978), magnetic fields, and also the average potential of the nuclei in solids, by using neutron mirrors or prisms. A possible fifth way is by neutron interferometry. This review will deal only with items (i), (ii) and (iii).

Time-of-flight methods

The new IN 6 time-of-flight spectrometer will use a new version of time focusing (Scherm et al. 1977a) to achieve improved resolution (20–100 μeV) at a chosen energy transfer and with increased neutron intensity. Three separate beams with different wavelengths fall upon the sample. The longest wavelength beam is chopped first and the shortest last in each cycle. Flight paths to the sample are arranged so that all pulses coincide at that point to give an effective short burst time.

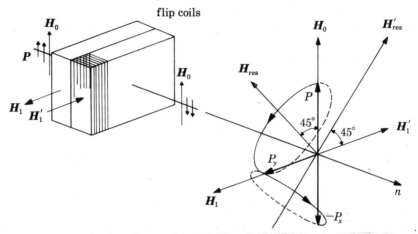

FIGURE 1. Continuous field spin flipper: the resulting H_{res} (or H'_{res}) from the constant magnetic fields H_0 and H_1 (or H'_1) is the axis inclined at 45° about which the spin P rotates when the neutrons pass through. By an appropriate choice of the time of flight in the field of the coils, a spin rotation from P_x to P_y (a single coil) or to $-P_x$ (the two coils) is obtained ($\frac{1}{2}\pi$ (or π-flipper); Badurek & Westphal 1975).

Another elegant time-of-flight method consists of bringing a polarized neutron beam to a spin analyser and inserting a spin flipper, as thin as possible, in front of the sample, which turns the spin through 180° when the electric current is switched on and off (Badurek & Westphal 1975) (figure 1). Such a non-mechanical device is limited by the inertia of the electrical switching process: rise times of ca. 4 μs were achieved with a beam of 5 cm × 5 cm. This method is already in use on the spectrometer D7. Shorter pulses would give shorter flight paths and thus greater intensity. Finally, the geometrical spread of the flight path due to sample and detector thickness sets a limit in time-of-flight experiments. The development of thin detectors such as the scintillation detectors under study for the Rutherford Laboratory Spallation Neutron Source promise further progress (Wroe 1978).

Backscattering spectroscopy

Diffraction at Bragg angles $\theta_B = 90°$ is well known as a method to achieve an extremely high resolution (Alefeld et al. 1969). In this backscattering situation, the neutron wavelength depends on θ_B only to second order. The remaining resolution width, caused by second-order effects in θ_B and by primary extinction in the crystal, is of the order of 0.2 μeV f.w.h.m. at best for an Si (111) reflexion (IN 10 spectrometer at the I.L.L.) and there is no way to pass far beyond this resolution limit.

A certain drawback of the existing backscattering instruments is their lack of flexibility. The range of energy transfers to be covered by the Doppler drive at the monochromator (piston of a motorcycle) is limited to transfers $|\hbar\omega| \leqslant 20$ μeV. Furthermore, there is no way to adapt the energy resolution to the problem under investigation. In this connection, a new instrument is under construction at the I.L.L. called IN 13 (Buevoz & Heideman 1978), which covers a much larger range of energy transfers $\hbar\omega$, and of resolutions (figure 2). By changing θ_B from exact backscattering to $\theta_B = 70°$ and changing the lattice parameter of the CaF_2 (422) monochromator by heating it up to 500 °C, an energy scan from -300 to $+500$ μeV, and an energy resolution in the range between 3 and 20 μeV, can be achieved.

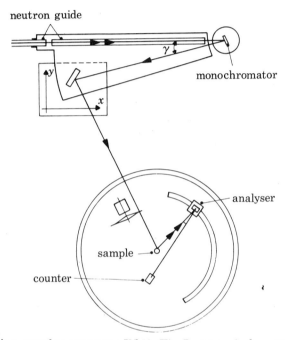

FIGURE 2. High-resolution crystal spectrometer IN 13. The Bragg angle $\theta_B = 90° - \gamma$ can be varied from 70° to 90° (backscattering), and the lattice plane distance of the monochromator by heating the crystal. Several analysers with associated detectors are provided. The energy scan is affected by varying γ and/or the monochromator temperature. The deflector crystal on the $x-y$ plate ensures that the beam always strikes the sample (Buevoz & Heidemann 1978).

The broadening of the resolution width increases the intensity, provided the crystal is sufficiently imperfect. As a future development, a furnace is under consideration that applies a heat gradient over the crystal, thus providing a controlled increase of the resolution width. This spectrometer, installed at a thermal guide, reaches Q values as high as 5.5 Å^{-1}†. This is important for many experiments dealing with diffusion in solids and with molecular rotations in crystals, since information on the geometry of such motions is only obtained if $1/Q$ is of the same magnitude as the typical distances of these motions.

A further development, existing only in a preliminary state, aims at the highest possible resolution (i.e. 0.3 μeV) in connection with very large energy transfers, $\hbar\omega$. This can be achieved by combining monochromators with different analyser crystals. By using a sufficiently fast Doppler drive (Bauer 1972), the energy gaps for different pairs of crystals can be covered

† 1 Å = 10^{-10} m = 10^{-1} nm.

entirely. A spectrometer of this kind would allow the investigation of low energy librational states up to nearly 1 meV (e.g. CaF_2 (200) paired with Ge (111) yielding 700 μeV $\leq \hbar\omega \leq$ 940 μeV; A. Heidemann & Jenkins, 1979), unpublished).

Spin echo spectroscopy

An elegant time-of-flight method developed in the I.L.L. uses the 'clock' individually attached to the neutron, namely the Larmor precession in magnetic field sections before and after the scattering sample. The principle of this spectrometer, called 'spin echo', has been described in detail (Mezei 1972, 1978; Hayter 1978), and I deal only with several relevant aspects of the method (see figure 3). The flight-time difference before and after scattering is determined by the difference of the Larmor angles in the magnetic field sections before and after the sample,

$$\delta\phi = (\Omega_0 L_0/v_0) - (\Omega_1 L_1/v_1); \tag{1}$$

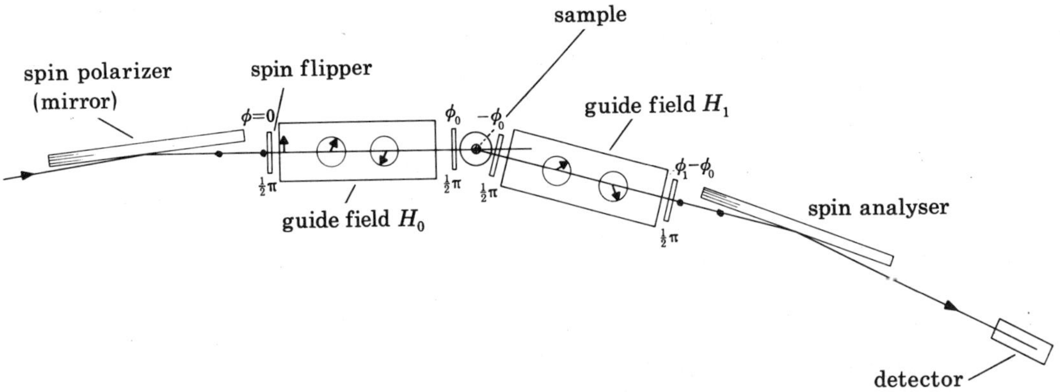

FIGURE 3. Principle of the spin echo spectrometer. A magnetized mirror or diffraction crystal can be envisaged as spin analyser or polarizer. The $\frac{1}{2}\pi$ coils operate on the principle shown in figure 2. The incident spin is perpendicular to the plane of the drawing. The difference of Larmor angles for the guide fields before and after scattering $\delta\phi$ is measured (Mezei 1972).

where $v_{0,1}$ and $L_{0,1}$ are the neutron velocity and the flight distance before and after scattering, respectively, and $\Omega = (4\pi\mu_n/h)H$ is the corresponding Larmor frequency. To avoid phase differences for neutrons entering or leaving the magnetic field sections at different positions in the beam, each neutron has to be rotated such that it starts with the same angle of precession within a well defined start (or arrival) plane (figure 3). This is achieved by a 90° spin flipper (action of only one of the two coils shown in figure 1). To obtain the flight-time *difference* according to (1), a spin rotation has to be applied near the sample position by a π-coil.

For purely elastic scattering ($\hbar\omega = 0$) and for an exactly symmetric arrangement ($H_0 L_0 = H_1 L_1$), one obviously has $\delta\phi = 0$. For a very small energy transfer,

$$\hbar\omega = \tfrac{1}{2}m/(v_0^2 - v_1^2) \ll \tfrac{1}{2}mv_0^2, \tag{2}$$

a change of the phase angle is observed which depends on $\hbar\omega$, and which is measured by means of the polarization behind the field section H_1. Differentiation of (1) and (2) with respect to v_1 and v_0 leads to the condition that $\delta\phi$ is *linear* in $\hbar\omega$, namely

$$H_0/v_0^3 = H_1/v_1^3. \tag{3}$$

This yields
$$\delta\phi = \phi_0(\hbar\omega/2E_0). \tag{4}$$

Since the polarization after the second spin flipper is proportional to cos $\delta\phi$, a quasi-elastic energy distribution of the scattered neutrons, $S(\omega)$, produces an average polarization

$$\langle P(\phi_0)\rangle = P_0 \int S(\omega) \cos(\phi_0 \hbar\omega/2E_0)\, d\omega, \tag{5}$$

where P_0 is the polarization for purely elastic scattering. Obviously, the average polarization as a function of ϕ_0 or of the magnetic field $H_1 = H_0$ directly yields the Fourier transform of $S(\omega)$. Equation (5) holds for incident spectra of a width below, say, 10%.

The spin echo spectrometer IN 11 based on this principle is now working in routine operation. To define its resolution, we assume that $S(\omega)$ is a Lorentzian with a half width 2Γ. Then the polarization ratio is given by $\langle P\rangle/P_0 = \exp(-\pi\Gamma N/E_0)$, where N is the total number of Larmor precessions (ca. 10^4) and E_0 is the incident energy. We define the resolution by a drop of $\langle P\rangle/P_0$ from 1.0 to 0.95 for the magnetic field range covered by the instrument (this value is considered as an easily measurable polarization change). This definition leads to an accuracy for the width of about 3×10^{-9} eV with 8 Å neutrons.

Each neutron obviously measures its velocity change individually (independent of v_0 to first order). Consequently, this very high resolution can be achieved with a broad incident spectrum, with the spectrometer IN 11, at present with a width of about 5×10^{-4} eV. This is 10^5 times larger than the resolution in energy transfer. In other words, the spin echo principle decouples the ω-resolution from the width of the incident spectrum. (This should be compared with the fact that the backscattering spectrometer decouples the ω resolution from the *angular width* of the incident beam.) Up to now, experience has demonstrated that the spin echo instrument IN 11 is certainly superior to the backscattering instrument IN 10 for high resolution quasi-elastic scattering experiments as long as they also require a good resolution in momentum (less than 10^{-2} Å^{-1}), in particular for critical scattering, or coherent polymer scattering. On the other hand, the backscattering spectrometer is normally equally good or superior for most of the incoherent scattering experiments.

At present, two further applications of the spin echo principle are under investigation, and will be tested at the D 10 spectrometer in the near future.

(i) The high resolution of the spin echo method allows a reduction of the thermal diffuse scattering (t.d.s.) occurring in the vicinity of Bragg reflexions (Hayter et al. 1979). This is feasible by separating the elastic part of the scattering, where the Larmor angle difference $\delta\phi$ is zero, from the phonon scattering, which leads to dephasing and therefore to a decay of the average polarization. A separation of phonons with energies not more than 50 μeV appears possible, which may reduce the t.d.s. contribution by a factor of more than 10. This will be important for accurate structure determinations.

(ii) The high resolution of the spin echo method can be applied to investigate the shape of phonon lines. Since the phonon peaks occur at relatively large energy transfers, this requires the asymmetric operation of a spin echo instrument, i.e. $v_0 \neq v_1$ and $H_0 L_0 \neq H_1 L_1$ in (1). Calculations have demonstrated that a triple-axis spectrometer with polarized neutrons applying this asymmetric spin echo principle will achieve a resolution which is about ten times better than the values achieved with triple-axis spectrometers like IN 3 with focusing, at a flux 30 times lower. The resolution with respect to the wavevector of the phonons, q, compatible with such an energy resolution had to be 10^{-4} to 10^{-5} Å^{-1}. To avoid the intensity loss caused by such a q-resolution, $q\omega$-focusing has to be applied. It can be achieved if the axis of the magnet is inclined against the beam direction so that a linear relation is obtained between the inclination of a neutron path in the magnet against the axis, and the phase angle $\delta\phi$ (Mezei 1978; Pynn 1978).

MULTIDETECTORS AND FOCUSING CRYSTALS TO INCREASE COUNT RATES

Multicell counters

The conventional method of investigating crystal diffraction patterns, or diffuse scattering from disordered crystals, is an intensity measurement point by point. Multicell counters yield an enormous gain in experimental time and/or statistical accuracy. This would hold, in particular, for crystals with large lattice cells, where many reflexions have to be studied. For the first time, multicounters were introduced in the I.L.L. for the small-angle cameras D 11 and D 17, using BF_3 (Allemand et al. 1975). More recently, a multicounter 'slice' for a single crystal diffractometer is under construction (D 19) filled with He^3 at 10 bar†. The counter has 16×512 cells for neutron localization, with a resolution of about $0.1° \times 0.2°$. This device is considered as a prototype for a bigger and advanced cylindrical multicounter for single crystal work (in particular for proteins) with a size of $70° \times 120°$, and about 10^5 cells. Such a counter would create data fluxes of 10^5 words per second, and, apart from mechanical difficulties with the counter itself, the on-line data reduction would be a serious problem.

A different application of multicell counters is the simultaneous registration of diffraction patterns as a function of time. This has many applications for kinetic experiments, as concerns problems of chemisorption, solid state reactions (as intercalation), or H–D exchange in macromolecules. A banana-shaped counter with 400 cells, covering an angular range of $80°$, is operated with the powder diffractometer D1B. The development of such detectors obviously aims at very high count rates. The best expected for the near future (detector D 20, ^{3}He, with 1800 wires) would be a count of about 10^5/s per cell. If 10^3 events per reflexion are considered sufficient in terms of statistics, a pattern could be measured within 10 ms. The use of conversion electrons for multicounters from neutron capture in ^{157}Gd is an alternative under development (Jeavons et al. 1978).

Increasing solid angles

In the past, it has been repeatedly argued that the virtue of thermal neutron guides is diminished by the small critical angle γ_c accepted by total reflexion. In particular, for the vertical direction, relatively large angles are tolerable and desirable in many instances (in particular for diffractometers). In this respect, considerable progress has been achieved by *vertical focusing*. This means 'transforming' the great height of a neutron guide into angular width. A typical *monochromator* for vertical focusing has a height of 10–15 cm, consisting of 10–30 lamellae made of Cu, Si, Ge or pyrolytic graphite, on an elastic metal sheet with variable radius of curvature. At a focal distance of say 2 m, the resulting intensity spot may have a height of about 3 cm, thus increasing the intensity by a factor of 3. The vertical height of the focal spot is essentially caused by the mosaic spread of the monochromator and the divergence of the incident beam. Careful cold working of ideal crystals with selected glide planes of dislocations leads to the desirable increase of spread in the horizontal plane, but, at the same time, leaves the mosaic spread small in the vertical direction (Freund 1975; Freund & Forsyth 1978).

† 1 bar = 10^5 Pa.

Such devices are used at the I.L.L. for triple-axis spectrometers and diffractometers. Recent developments of focusing in space *and* in energy by horizontally curved crystals, applied in triple axis spectrometry will not be treated here (Scherm *et al.* 1977b).

An interesting way to improve neutron mirrors was developed recently by Mezei (1976, 1977). A multilayer sheet can be produced by repetitive evaporation of two materials with a different refractive index. Consequently, in addition to normal total reflexion below the

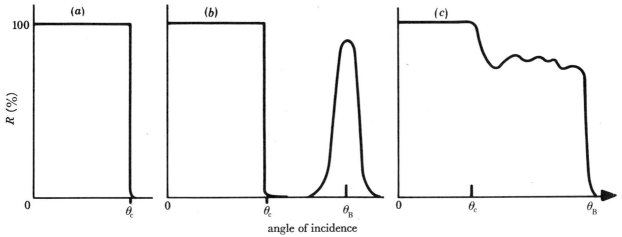

FIGURE 4. Neutron reflectivity, R (schematic): (a) a simple total reflecting mirror, (b) a multilayer mirror (equidistant double layers of different refractive index) and (c) variable distance between layers. θ_c = critical angle, θ_B = Bragg or Bragg cut-off angle (from Mezei 1976, 1977). For a Ni-coated guide at 2 Å, $2\theta_c = 0.40°$ (for ^{58}Ni, $2\theta_c = 0.47°$). For the multilayer mirror, θ_c corresponds to the average refractive index.

critical angle $\bar{\gamma}_c$ of the mixture of the two materials, a Bragg reflexion due to the periodic structure appears at an angle beyond $\bar{\gamma}_c$. By introducing a gradient with respect to the layer distance, and by a proper choice of this gradient, it is possible to broaden the Bragg peak asymmetrically until it merges with the cut-off angle $\bar{\gamma}_c$ ('supermirrors', figure 4). By using an alternation between magnetic and non-magnetic layers, this type of mirror also serves as a spin polarizer. So far, the following characteristics have been achieved by F. Mezei: 75 double layers Fe/Ag with a gradient of distance (average 300 Å) yields a reflectivity of nearly 100 % up to $\bar{\gamma}_c$, which decreases to about 70 % between $\bar{\gamma}_c$ and the Bragg cut-off angle, which occurs at about 2.4 $\bar{\gamma}_c$. Above $\bar{\gamma}_c$, the polarization of a reflected neutron beam is 99 %; below $\bar{\gamma}_c$ it decreases to 97 %. Obviously these mirrors are excellent polarizers and are already used in several instances. As concerns their application for neutron guides, the gain by the large cut-off angle is only effective if the number of reflexions is small, because of the poor reflectivity above $\bar{\gamma}_c$. An evaporation unit for routine production of such mirrors will start operating at the end of 1979.

3. Production of ultra-cold neutrons

This topic, being somewhat outside the main scope of the review, is related to experiments with ultra-cold neutrons (u.c.n.), where 'ultra-cold' means kinetic energies of the order of 10^{-7} eV ($\lambda = 1000$ Å, $V = 4$ m/s). This is below the average nuclear interaction potential of most solids (also below the interaction potential of the neutron dipole moment with high magnetic fields). Consequently, total reflexion can be achieved even at perpendicular incidence, which is the basis of many interesting experiments, for instance the measurement of the electric dipole moment of the neutron (or of its lifetime against β decay). This paper does not deal with such experiments and I restrict myself to the problem of the 'production' of such neutrons (for reviews see Lushikov 1977; Steyerl 1977; Golub & Pendlebury 1979).

The neutron current $I(E)$ at energy E which leaves a neutron guide at the reactor is described by a Maxwellian at the moderator temperature T_M. This is a good approximation for a D_2O moderator, even at very low energies. One gets

$$I(E)/dE = \phi_0(\Delta\Omega/4\pi) \exp(-E/kT_M) \, EdE/(kT)^2 \, dE, \qquad (6)$$

where $\Delta\Omega/4\pi \approx \frac{1}{4}$ is the solid angle for u.c.n. accepted by the guide, and ϕ_0 is the total thermal flux.

Integration of (6) up to, for our example, $E = 1.9 \times 10^{-7}$ eV (6.2 m/s) yields a total extracted u.c.n. current of $I_{u.c.n.}$ ($\leqslant 6.7$ m/s) $= (\frac{1}{8}) \phi_0(E_{u.c.n.}/kT)^2 \approx 3 \times 10^3$ s^{-1} cm^{-2} with $\phi_0 = 0.4 \times 10^{15}$ s^{-1} cm^{-2} for the disturbed thermal flux, as obtained at the inclined hole IH 3 (Ageron 1978).

Owing to high absorption in the wall of the beam-hole thimble, only a small fraction of the u.c.n. could escape. To improve this situation, a special installation has been constructed in the IH 3 beam hole, called 'PN 5'. The front wall of the tube has been covered on its inside by a *ca.* 2 mm layer of streaming water, with a 100 μm aluminium window to prevent absorption. The H_2O scatters the thermal neutrons into the u.c.n. region (originally depleted by absorption); thus the spectrum in this range recovers. Nevertheless, owing to reflexion losses in the u.c.n. neutron guide (10 m long, 7 cm × 7 cm cross section, Ni-coated glass), and owing to air gap and window losses, only about 80 u.c.n. s^{-1} cm^{-2} reach the detector (Ageron 1978).

A method of improving the extraction losses is a neutron turbine (Steyerl 1975) sitting outside the reactor at the end of a neutron guide. In momentum space, it shifts the neutrons from high energies down to u.c.n. energies, thus avoiding the extraction losses at u.c.n. energies (of course, Liouville's theorem says that one certainly cannot get more than the density corresponding to the equilibrium value in momentum space, $\rho = d^6N/dx \, dy \, dz \, dp_x dp_y dp_z = \{\phi_0/8\pi m_N(kT)^2\} \exp(-E/kT)$; see, for example, Maier-Leibintz 1966). This method has been worked out at the Munich reactor and its application at the I.L.L. is being considered.

In many experiments, a high u.c.n. *density*, $\rho_{u.c.n.}$, is needed instead of a high current. Outside the 'PN 5' hole, the u.c.n. current quoted above would lead to a value

$$\rho_{u.c.n.} = 4I_{u.c.n.}/\bar{V}_{u.c.n.} = 0.8/\text{cm}^3 \qquad (7)$$

for an ideal and empty bottle, with $\bar{V}_{u.c.n.} = 4$ m/s.

Some time ago a very interesting proposal was developed to increase considerably the u.c.n. density outside the reactor (Golub & Pendlebury 1975, 1979). A beam of 'medium' energy (*ca.* 10^{-3} eV), with no appreciable extraction losses, penetrates the wall of a cryostat filled with superfluid ^{4}He. There the neutrons interact with the phonon-like elementary excitations of the helium. A large proportion of the neutrons lose almost all of their energy and momentum in these processes, with the simultaneous production of quanta of helium elementary excitations. (For a *solid* at extremely low temperature this would not apply: although there exist phonon excitations, at low temperature the majority of the scattering processes are purely elastic and therefore not effective.) For a 3 m long closed helium 'neutron bottle', a stationary density can be produced, determined by the equilibrium between the u.c.n. production rate per volume unit, $P_{u.c.n.}$, and the unavoidable losses. The losses are described by an effective lifetime τ_{eff}. This results primarily from the temperature-dependent energy gain or up-scattering processes in superfluid helium (τ_g), from the walls (τ_w), from the absorption in the isotope ^{3}He (τ_a, avoidable by isotope purification), and of course from the β-decay of the neutron $\tau_\beta = 907$ s. For the equilibrium density in the superfluid helium one obtains

$$\rho_{u.c.n.} = P_{u.c.n.}/\tau_{eff} = P_{u.c.n.}(1/\tau_g + 1/\tau_w + 1/\tau_a + 1/\tau_\beta + \ldots). \tag{10}$$

Under favourable conditions a lifetime is expected that is governed by the wall losses and the helium up-scattering; with $\tau_{eff} = 30\text{–}100$ s and with $P_{u.c.n.}$ from the approximately calculated helium up-scattering processes (Golub 1979), we obtain a density $\rho_{u.c.n.}$ for this bottle, which is almost 100 times greater than the figure obtained by using the existing 'PN 5' beam.

A preliminary test with a stationary beam, with incident neutrons of 10^{-3} eV in helium at 1.1 K, showed that the estimated production rate $P_{u.c.n.}$ is, in fact, achieved (Ageron et al. 1978). The loss processes due to up-scattering in helium (Golub 1979) and particularly in hydrogen absorbed on the walls (Stoika et al. 1978) can be estimated theoretically. It will certainly be possible to achieve improvements as regards the wall losses. Such a He facility is planned and it will essentially allow a repetition of the search for the neutron's electric dipole moment, and other experiments.

References (Springer)

Ageron, P. 1978 Internal I.L.L. report.
Ageron, P., Mampe, W., Golub, R. & Pendlebury, J. M. 1978 *Physics Lett.* A **66**, 469.
Alefeld, B. Birr, M. & Heidemann, A. 1969 *Naturwissenschaften* **56**, 410.
Allemand, R., Bourdel, J., Roudant, E., Convert, P., Ibel, K., Jacobé, J. Cotton, J. P. & Farnoux, B. 1975 *Nucl. Instrum. Methods* **126**, 29.
Badurek, G. & Westphahl, G. P. 1975 *Nucl. Instrum. Methods* **128**, 315.
Bauer, G. 1972 *Kerntechnik Atompraxis* **14**, 82.
Buevoz, J. L. & Heidemann, A. 1978 Internal I.L.L. report.
Freund, A. 1975 *Nucl. Instrum. Methods* **124**, 93.
Freund, A., Forsyth J. B. 1978 In *Neutron scattering in materials science* (ed. G. Kostorz). Academic Press.
Golub, R. 1979 Internal I.L.L. report.
Golub, R. & Pendlebury, J. M. 1975 *Physics Lett.* A **53**, 133.
Golub, R. & Pendlebury, J. M. 1979 *Rep. Prog. Phys.* **42**, 439.
Hayter, J. B. 1978 *Z. Phys.* B **31**, 117.
Hayter, J. B., Lehman, M. S., Mezei, F. & Zeyen, C. M. E. 1979 *Acta Crystallogr.* A **35**, 333.
Jeavons, A. P., Ford, N. L., Lindberg, B. & Sachat, R. 1978 *Nucl. Instrum. Methods* **148**, 29.
Luschikov, V. I. 1977 *Physics Today*, June, p. 42.
Maier-Leibnitz, H. 1966 *Nukleonik* **8**, 61.
Mezei, F. 1972 *Z. Phys.* **255**, 146.
Mezei, F. 1976 *Communs Phys.* **1**, 81.
Mezei, F. 1977 *Communs Phys.* **2**, 41.
Mezei, F. 1978 In *Neutron inelastic scattering*, vol. 1, p. 125. Vienna: I.A.E.A.
Pynn, R. 1978 *J. Phys.* E **11**, 1133.
Scherm, R., Dianoux, A. J. & White, J. 1977a Internal I.L.L. report.
Scherm, R., Dolling, G., Ritter, R., Schedler, E., Teuchert, W. & Wagner, V. 1977b *Nucl. Instrum. Methods* **143**, 77.
Steyerl, A. 1975 *Nucl. Instrum. Methods* **125**, 461.
Steyerl, A. 1977 In *Springer Tracts mod. Phys.* **80**, 57.
Steyerl, A. 1978 *Z. Phys.* B **30**, 231.
Stoika, A. D., Stelkov, A. V. & Hetzelt, M. 1978 *Z. Phys.* B **29**, 349.
Wroe, H. 1978 Rutherford Laboratory Report.

INDEXES TO VOLUME 290 (B)

Author index

Anstis, S. M. The perception of apparent movement, 153.
Axe, J. D. Incommensurate structures, 593.

Barlow, H. B. The absolute efficiency of perceptual decisions, 71.
Benjamin, P. R., Slade, Carole T. & Soffe, S. R. The morphology of neurosecretory neurones in the pond snail, *Lymnaea stagnalis*, by the injection of Procion Yellow and horseradish peroxidase, 449.
Bentley, G. A. & Mason, S. A. Neutron diffraction studies of proteins, 505.
Bentley, G. A. *See* Finch, Lewit-Bentley, Bentley, Roth & Timmins.
Blight, A. R. & Llinás, R. The non-impulsive stretch-receptor complex of the crab: a study of depolarization–release coupling at a tonic sensorimotor synapse, 219.
Bomchil, G., Hüller, A., Rayment, T., Roser, S. J., Smalley, M. V., Thomas, R. K. & White, J. W. The structure and dynamics of methane adsorbed on graphite, 537.
Braddick, O. J. Low-level and high-level processes in apparent motion, 137.
Brown, P. Jane, Forsyth, J. B. & Mason, R. Magnetization densities and electronic states in crystals, 481.

Campbell, F. W. The physics of visual perception, 5.
Chater, K. F. *See* Hopwood & Chater.
Clarke, Patricia H. Microbiology and pollution: the biodegradation of natural and synthetic organic compounds, 355.
Coltheart, M. The persistences of vision, 57.
Cowley, R. A. Percolation in antiferromagnetic insulators, 583.
Cox, S. *See* Leslie, Jenkin, Hayter, White, Cox & Warner.

Davies, D. S. Industrial microbiology: a view from Whitehall, 281.
Dunnill, P. Immobilized cell and enzyme technology, 409.

Enderby, J. E. Neutron diffraction, isotopic substitution and the structure of aqueous solutions, 553.

Fender, B. E. F., Hobbis, L. C. W. & Manning, G. The U.K. Spallation Neutron Source, 657.
Finch, J. T., Lewit-Bentley, A., Bentley, G. A., Roth, M. & Timmins, P. A. Neutron diffraction from crystals of nucleosome core particles, 635.
Forsyth, J. B. *See* Brown, Forsyth & Mason.
Frisby, J. P. & Mayhew, J. E. W. Spatial frequency tuned channels: implications for structure and function from psychophysical and computational studies of stereopsis, 95.
Frost, J. C., Leadbetter, A. J. & Richardson, R. M. Molecular crystals and liquid crystals: new results for *t*-butyl chloride, 567.

Georgeson, M. A. Spatial frequency analysis in early visual processing, 11.
Gregory, R. L. Perceptions as hypotheses, 181.

Haddock, B. A. Microbial energetics, 329.
Hartley, B. S. Introductory remarks to a Discussion on industrial microbiology, 279.
Hayter, J. B. *See* Leslie, Jenkin, Hayter, White, Cox & Warner.
Hobbis, L. C. W. *See* Fender, Hobbis & Manning.
Hopwood, D. A. & Chater, K. F. Fresh approaches to antibiotic production, 313.
Hughes, J. Biologically active peptides: prospects for drug development, 387.
Hüller, A. *See* Bomchil, Hüller, Rayment, Roser, Smalley and others.

Jacrot, B. The use of neutrons to study protein–RNA interactions, 627.
Jenkin, G. T. *See* Leslie, Jenkin, Hayter, White, Cox & Warner.
Julesz, B. Spatial nonlinearities in the instantaneous perception of textures with identical power spectra, 83.

Leadbetter, A. J. *See* Frost, Leadbetter & Richardson.
Lee, D. N. The optic flow field: the foundation of vision, 169.
Lennie, P. Perceptual signs of parallel pathways, 23.
Leslie, M., Jenkin, G. T., Hayter, J. B., White, J. W., Cox, S. & Warner, G. Precise location of hydrogen atoms in complicated structures by diffraction of polarized neutrons from dynamically polarized nuclei, 497.
Lewit-Bentley, A. *See* Finch, Lewit-Bentley, Bentley, Roth & Timmins.
Llinás, R. *See* Blight & Llinás.
Longuet-Higgins, H. C. *See* Sutherland (N. S.) & Longuet-Higgins.

Manning, G. *See* Fender, Hobbis & Manning.
Manton, Irene, Sutherland, J. & Oates, K. A reinvestigation of collared flagellates in the genus *Bicosta* Leadbeater with special reference to correlations with climate, 431.
Marmion, B. P. Prospects for new viral vaccines, 395.
Marr, D. Visual information processing: the structure and creation of visual representations, 199.
Mason, R. *See* Brown, Forsyth & Mason.
Mason, S. A. *See* Bentley & Mason.
Mayhew, J. E. W. *See* Frisby & Mayhew.
Middendorf, H. D. & Randall, Sir John. Molecular dynamics of hydrated proteins, 639.
Mitchell, E. W. J. & Stewart, R. J. Diffuse neutron scattering from crystal imperfections, 511.
Morgan, M. J. Analogue models of motion perception, 117.
Moulden, B. After-effects and the integration of patterns of neural activity within a channel, 39.
Murray, K. Genetic engineering: possibilities and prospects for its application in industrial microbiology, 369.

Oates, K. *See* Manton, Sutherland & Oates.

Pelizzari, C. A. *See* Sköld & Pelizzari.
Pendlebury, J. M. & Smith, K. The electric and magnetic moments of the neutron, 617.
Postgate, J. R. Prospects for the exploitation of biological nitrogen fixation, 421.

Randall, Sir John. *See* Middendorf & Randall.
Rayment, T. *See* Bomchil, Hüller, Rayment, Roser, Smalley and others.
Richardson, R. M. *See* Frost, Leadbetter & Richardson.
Righelato, R. C. Anaerobic fermentation: alcohol production, 303.
Roser, S. J. *See* Bomchil, Hüller, Rayment, Roser, Smalley and others.
Roth, M. *See* Finch, Lewit-Bentley, Bentley, Roth & Timmins.

Schindler, P. Enzyme inhibitors of microbial origin, 291.
Schmatz, W. Magnetic diffuse and small-angle scattering, 527.
Sköld, K. & Pelizzari, C. A. Elementary excitations in liquid ^{3}He, 605.
Slade, Carole T. *See* Benjamin, Slade & Soffe.
Smalley, M. V. *See* Bomchil, Hüller, Rayment, Roser, Smalley and others.
Smith, K. *See* Pendlebury & Smith.
Smith, S. R. L. Single cell protein, 341.
Soffe, S. R. *See* Benjamin, Slade & Soffe.
Springer, T. Developments in experimental neutron physics at the Institut Laue–Langevin, 673.
Stewart, R. J. *See* Mitchell & Stewart.
Sutherland, J. *See* Manton, Sutherland & Oates.
Sutherland, N. S. & Longuet-Higgins, H. C. Introduction to a Discussion on the psychology of vision, 3.

Thomas, R. K. *See* Bomchil, Hüller, Rayment, Roser, Smalley and others.
Timmins, P. A. *See* Finch, Lewit-Bentley, Bentley, Roth & Timmins.

Warner, G. *See* Leslie, Jenkin, Hayter, White, Cox & Warner.
White, J. W. *See* Bomchil, Hüller, Rayment, Roser, Smalley and others; *also* Leslie, Jenkin, Hayter, White, Cox & Warner.

Subject index

Bicosta: climatic effects, 431.

Climate and morphology in *Bicosta*, 431.

Discussion on neutron scattering in biology, chemistry and physics; for detailed contents see pp. 479 and 480.
Discussion on new horizons in industrial microbiology; for detailed contents see pp. 277 and 278.
Discussion on the psychology of vision; for detailed contents see pp. 1 and 2.

Industrial microbiology, Discussion; for detailed contents see pp. 277 and 278.

Neurosecretory neurones, 449.
Neutron scattering in biology, chemistry and physics, Discussion; for detailed contents see pp. 479 and 480.
Non-impulsive neurons, 219.

Psychology of vision, Discussion; for detailed contents see pp. 1 and 2.

Snail neuronal morphology, 449.
Synaptic transmission, 219.

End of the two hundred and ninetieth volume – Series B